Other Books by Lem Moyé

- *Statistical Reasoning in Medicine: The Intuitive P-Value Primer*
- *Difference Equations with Public Health Applications* (with Asha S. Kapadia)
- *Multiple Analyses in Clinical Trials: Fundamentals for Investigators*
- *Probability and Statistical Inference: Applications, Computations, and Solutions* (with Asha S. Kapadia and Wen Chan)
- *Statistical Monitoring of Clinical Trials: Fundamentals for Investigators*
- *Statistical Reasoning in Medicine: The Intuitive P-Value Primer— 2nd Edition*
- *Face to Face with Katrina's Survivors: A First Responder's Tribute*
- *Elementary Bayesian Biostatistics*
- *Saving Grace* (A Novel)

Weighing the Evidence

Duality, Set, & Measure Theory in
Clinical Research Analyses

LEM MOYÉ

Order this book online at www.trafford.com
or email orders@trafford.com

Most Trafford titles are also available at major online book retailers.

© Copyright 2020 Lem Moyé.

All rights reserved. No part of this publication may be reproduced, stored in a retrieval system, or transmitted, in any form or by any means, electronic, mechanical, photocopying, recording, or otherwise, without the written prior permission of the author.

Print information available on the last page.

ISBN: 978-1-4907-9975-9 (sc)
ISBN: 978-1-4907-9977-3 (hc)
ISBN: 978-1-4907-9976-6 (e)

Library of Congress Control Number: 2020903038

Because of the dynamic nature of the Internet, any web addresses or links contained in this book may have changed since publication and may no longer be valid. The views expressed in this work are solely those of the author and do not necessarily reflect the views of the publisher, and the publisher hereby disclaims any responsibility for them.

Any people depicted in stock imagery provided by Getty Images are models, and such images are being used for illustrative purposes only.
Certain stock imagery © Getty Images.

Trafford rev. 03/06/2020

Trafford PUBLISHING www.trafford.com

North America & international
toll-free: 1 888 232 4444 (USA & Canada)
fax: 812 355 4082

Contents

Preface xiii
Introduction xvii

The Ten-Thousand-Foot View... 1
    What Is Duality? 3
    Plausible versus Confidence Intervals 5
    This Book's Methodology—Parse, Channel, and Accumulate 5
    Set Theory and Quanta Analysis... 6

Statistical Theology... 9
    The Two-Minute Problem 10
    The Cutting Room Floor 11
    Two Trajectories..11
    Epidemiology and Health Care Delivery...................................... 12
    Probability, Statistics, and Prediction 13
    Eruption 14
    The 1920s and the Appearance of Statistical Significance 16
    Observational Scientists React 17
    Enter the Administrators..19
    Rule of Thumb 21
    Transmogrification 22
    Pushback 23
    Doubling Down on the P-value 23
    Clinical Research Complexities... 25

| Wasteland 27
| Reproducibility 29
| Conclusions 32

What Do We Require of This New Approach?................................39

The Basics of Set Theory...43
    Motivation for This Work.. 43
    What Are Sets? 44
    Introducing Relationships between Sets 44
    Set Operations 45
    Venn Diagrams... 48
    Distribution Law of Sets 50
    De Morgan's Law.. 50
    Set Generation and σ-Algebras 51
    Why We Need σ-Algebras 54

Elementary, Set, and Measureable Functions.......................................55
    Measurability.. 56
    Broadening the Elementary Function... 58
    Summary... 59

Measure and Its Properties 61
    Elementary Path 62
    What Is Measure Theory 63
    Accumulation 63
    Notation 69

Working with Measure's First Three Properties 71
    Review of the Sample Space and Sigma Algebras 71
    Measure versus Measurable Functions—Properties of Measure..... 72
    Measure of the Union of Two Sets 74
    Summary... 79

Property 4 of Measure: Countability 81

An Interlude 85

Functions and Measures on Analysis Regions.....................................89
    What Constitutes an Analysis?.. 89
    Regions of Analyses 91

Defining the Content of an Analysis	93
The Content of an Analysis	94
Analysis Redundancy	97
Computing the Content of Analysis Unions	98
Converting ψ-Content to ψ-Measure	103
Chapter Summary	112
Measuring Analysis Sets (Quanta Analysis)	113
Computing Quanta Sums	114
Strategy in Calculating Quanta	116
A Breather …	119
Some Helpful Observations	120
A First Demonstration	123
Example—A Single Primary Endpoint	124
Initial Analysis Sequencing Observation	127
Analysis Priorities and Quanta Paths	129
Sequencing Variant Quanta Values	129
Assigning Location to Sequence Variant Analyses	130
Example—Multiple Primary Outcomes	132
At What Level Does Averaging Take Place?	134
Multiple Manuscripts	134
Subgroup Evaluations	136
Notation	138
Chapter Summary	138
Topside Functions	141
Return to Duality	142
Interval Parsing, Channeling, and Accumulating	143
Our Initial Concerns	144
Beginning Construction of the Plausible Interval	144
Setting the Bounds for the Plausible Interval	145
Parsing the Plausible Interval	147
What Would We Like from a Benefit Function?	148
Measurable Functions of Benefit and Harm	150

Putting It All Together 153
    Example 1: One and Only One Outcome—No Effect Size 156
    Example 2: One Outcome—Moderate Effect Size 158
    Example 3: One Outcome—Large Effect Size 159
    Example 4: One Outcome—Overwhelming Harm Effect 160
    Conclusions from Single Outcome Examples 161
    Example 5: Two Outcomes with Reversed Effects 161
    Example 6: Three Outcomes Each with Small Effects 164
    Example 7: Three Outcomes with One Disparity 166

Quanta Analyses and the Supremacy of Safety 169
    The Safety Disconnect in Research 170
    Safety Findings and Type I Error 170
    What Else Can We Do? 171
    Example—Heart Failure Therapy and Creatinine 172
    Summary 174

Managing Correlation Between Variables 175
    Proposed Formulation Using Determinants 176
    Correlations and Unions of Analyses 176
    Example 1—Regression Analysis Families 177
    Summary 179

Incorporating Exploratory Analyses 181
    Exactly What Are Exploratory Analyses? 181
    The Problem with Exploratory Analyses 183
    Does that Mean They Should Not Be Published? 185
    Duality and Quanta Analyses in Exploratory Evaluations 185

Contributions of Other Measurable Functions 189

Limitations 193
    The Quanta Measure 193
    The Topside Function Is Not Optimal 193
    The Methodology Function Is Not Unique 194
    There Is No Sample Size Formula 195
    Lack of Independent Confirmation 196
    A Real Work Test Is Lacking 196

Conclusions—Queen Anne's Decree 197
    Quanta Analysis 197
    The Need for a Solid Research Foundation Endures 198
    Cultural Conflict of Interest 198
    Taking Matters into our Hands 199
    Longitude 200

Biographies 203
    Georg Cantor 205
    Bernhard Riemann 209
    Henri Lebesgue 211
    Thomas Joannes Stieltjes 215
    Andrey Kolmogorov 217

Index 221

To Dixie and the DELTS

Preface

"How would I analyze a clinical trial if there was no such thing as statistical hypothesis testing?"

Before we get to that, let's take a moment to get acquainted.

I became a physician in 1978, earned my PhD in biostatistics and epidemiology in 1987, and immediately began a thirty-two-year career as a faculty member at the University of Texas School of Public Health in Houston, Texas.

That was the heyday of clinical trials.

Public health was a well-organized discipline, receiving full support from the scientific community and the National Institutes of Health. Its scientists, working in the government as well as the private sector, were committed to rubbing out chronic disease with the same energy and zeal that helped eradicate or reduce the prevalence of many infectious diseases. My particular focus was cardiology, where the one-two punch of heart attacks and resultant heart failure stole the lives of millions of Americans each year.

Biostatistics was an important instrument in the toolkit of clinical investigation, and I immersed myself fully. However, while I have always enjoyed the background mathematics, I became concerned about its application to clinical research.

The use of estimation theory was exquisite; I have no issue with the computations of estimators e.g., means, event rates, hazard ratios, or other quantities based on estimation theory.

As a physician, it was the inference component that wore at me.

Statistical hypothesis testing, with its dichotomous decision metric (rejection or non-rejection of the null hypothesis) was ill-suited for clinical research. This should perhaps not be a surprise because, after all, it was not developed for health care research. The stunner is that despite this poor fit, statistical hypothesis testing has become an increasingly truculent tail, unmercifully wagging the clinical trial dog.

My first book, *The P-Value Primer*, attempted to sort out the proper role of biostatistics in clinical research. Nevertheless, investigators continued to lose ground to *p*-value primacy, ironically during a time of new confusion about what this quantity actually means.

Consider that while eminent statisticians have now called for reducing the threshold of statistical significance from 0.05 to 0.005,[1] the American Statistical Association, for the first time in its 177-year history, felt compelled to issue a statement clarifying for its own membership what *p*-values mean and how they should be used, a clarification that itself had to be explained.[2] Even the lay press has become involved, as the New York Times tried to unravel the mystery of *p*-value interpretation for its readers.[3]

How can it be that approximately ninety-five years after Ronald Fisher's first writings on statistical inference, statisticians and clinical investigators suffer continued confusion about its interpretation as the *p*-value dives deeper into clinical research?

Clinical researchers are, by and large, the victims of these insidious infiltrations of nebulous and confusing statistical inference procedures. Why do physician-scientists—otherwise so punctilious about clinical measures, e.g., a slight alteration in the movement of the left ventricular wall as determined by magnetic resonance imaging, or a small change in a cancer patient's biomarker level —willingly turn over their data to statistical hypothesis testing with its continued, slurring confusion over the interpretation of *p*-values?

It is a provocative question with a simple answer: they do it because they are told that they must. Mentors and department chairs, journal reviewers and editors, National Institutes of Health grant administrators, and the US federal Food and Drug Administration almost all de jure or de facto require statistical hypothesis testing.

So, a different tack is in order. Rather than just expatiate these issues, let's ask the question, "How would we analyze data from a clinical research effort if there was no such thing as statistical hypothesis testing?"

This provocative question instantiated a four-year quest on my part, leading to the development of a new construct and new quantitative tools. These ideas involve a concept termed *duality* and draw on the topics of set and measure theory from mathematics.

The purpose of this text is to expound on each of these topics and demonstrate that their application to clinical research provides new insight and addresses interpretative conundrums that statistical hypothesis testing cannot.

The audience for this book is broad; clinical trial researchers, biostatisticians, epidemiologists, and of course, students of these disciplines. This is a wide swath of expertise, and I have worked to use language to guide this audience, with it heterogeneous mathematics background though this book's technical chapters.

OK. Let's crack on.

References

[1] Benjamin, D. J., J. O. Berger, M. Johannesson, B. A. Nosek, E. J. Wagenmakers, R. Berk, K. A. Bollen, B. Brembs, L. Brown, C. Camerer, D. Cesarini, C. D. Chambers, M. Clyde, T. D. Cook, P. De Boeck, Z. Dienes, A. Dreber, K. Easwaran, C. Efferson, E. Fehr, F. Fidler, A. P. Field, M. Forster, E. George, R. Gonzalez, S. Goodman, E. Green, D. P. Green, A. G. Greenwald, J. D. Hadfield, L. V. Hedges, L. Held, T. Hua Ho, H. Hoijtink, D. J. Hruschka, K. Imai, G. Imbens, J. P. A. Ioannidis, M. Jeon, J. H. Jones, M. Kirchler, D. Laibson, J. List, R. Little, A. Lupia, E. Machery, S. E. Maxwell, M. McCarthy, D. A. Moore, S. L. Morgan, M. Munafó, S. Nakagawa, B. Nyhan, T. H. Parker, L. Pericchi, M. Perugini, J. Rouder, J. Rousseau, V. Savalei, F. D. Schönbrodt, T. Sellke, B. Sinclair, D. Tingley, T. Van Zandt, S. Vazire, D. J. Watts, C. Winship, R. L. Wolpert, Y. Xie, C. Young, J. Zinman, V. E. Johnson. "Redefine Statistical Significance." *Nat Hum Behav* 2, no. 1 (January 2018):6–10. https://doi.org/10.1038/s41562-017-0189-z.

[2] Wasserstein, R. L. and N. A. Lazar. "The ASA's Statement on p-values: Context, Process, and Purpose." *Am Stat* (2016). Epub ahead of print. https://doi.org/10.1080/00031305.2016.1154108.

[3] Kalaichandranb A. Worried About That New Medical Study? Read this First. Health & Wellness. Feb 5, 2020.

Introduction

The thesis of this book is that statistical hypothesis testing does not answer the actual questions that the clinical researcher has posed but instead answers a question that the researcher (1) has not asked and (2) has no interest in its answer.

My approach to this dilemma is to answer the question "If we only had estimation theory and not statistical hypothesis testing, how would we analyze clinical research data?" The resulting approach gives clinical researchers direct access to the answers to their fundamental research question: "Does the research help my patients or injure them?"

This book begins quite nonmathematically, discussing the philosophical concerns about the use of the p-value and the acculturation of generations of health care researchers to the use of statistical hypothesis testing even though it was not designed for clinical research from its first principles. Its inculcation has led to the institutionalization of physicians, biostatisticians, and administrators who frankly would be lost without this single number's presence.*

The clinical research community has permitted itself to be caught up in the tidal drift generated by the need for a computational, interpretative tool. While this device, in addition to the requirement of a disciplined, scientific protocol, added structure to research interpretation in the 1950s, it has, in my view, placed restrictions on research design that have nothing to do with biology, pathophysiology, or even logistics but is instead driven by the need to generate a p-value–based assessment of the impact of the intervention or exposure.

---

* Those readers who are already familiar with this dialogue can skip the first chapter 1.

This is not a conspiracy theory book. None of the *p*-value history that I provide is nefarious. While there have been experienced and prominent members of the statistical community who reinforced *p*-value primacy, there is no statistical-hypothesis-testing Darth Vader. In fact, many statisticians conduct statistical hypothesis testing because simply (1) that is what has been asked of them and (2) they know of no acceptable alternative. We all have ourselves to blame for this confused miasma. Our answer does not reside in a *Star Wars* overlord but in Shakespeare's *Julius Caesar*.

So, enough about the blame. Let's fix the problem.

The book combines a new approach (duality theory) with a well-established approach in mathematics (measure theory) to weigh the evidence in a clinical research effort supporting benefit and supporting harm. Duality theory states that an estimator of an effect in a clinical trial, be it a difference in mean change in diastolic blood pressure or a prevalence ratio, simultaneously contains evidence of benefit and evidence of harm. The evidence for each is extracted.

However, since multiple analyses from the same trial commonly utilize overlapping sets of observations and variables, the redundancy should be quantified and identified. The new developed tool (called quanta analysis) has its foundation in the basics of set and measure theory.

The combination of duality, set, and measure theory appears to be new.

The mathematics of measure theory is commonly taught at advanced levels. It need not always be so and this book is one of the exceptions.

The first examples offered by this book are almost absurdly simple yet are necessary for the reader to begin to gain some experience and intuition in the use of duality/quanta analyses. As the examples increase in sophistication, the reader can see how duality/quanta analysis assembles risk and benefits in increasingly complex clinical research scenarios.

In the end, the reader will know the theory and operation of this process as well as its strengths and weaknesses. I finish with some additional embellishments that can be useful—even illuminating—if pursued. Clinical trials are mute on these latter issues because there is little methodology to support their inclusion. However, they are of long-standing investigator interest.

This is my work, so I and I alone am responsible for any and all errors. Fortunately, it is easy to publish new book editions; so should mistakes slip into my writing, please point them out, and I will correct and release updated versions. This work, like life, is a work in progress and requires midcourse adjustments and corrections.

Finally, I have relied on many teachers and workers as my ideas have developed, going back to my days of driving frigid roads through barren winter landscapes to attend advanced probability courses at Purdue University. Barry Davis at the University of Texas School of Public Health has been a colleague and friend for over thirty years before I retired. I have learned from him at the Coordinating Center for Clinical Trials there. Robert Hardy, Mort Hawkins, and Asha Kapadia were fine mentors. Special thanks goes to Hulin Wu, Deijan Lai, and Hongjian Zhu, conversations with whom about the practical use of measure theory were particularly productive.

Ray Lipicky at the FDA was always pushing me to think anew. He always found room in my complacent bonnet to add yet one more bee.

My colleagues and mentors in the SHEP, SAVE, ALLHAT, CARE, SPOTRIAS, and CCTRN trials have each pulled at my intellectual gravity center, adjusting the trajectory of my thoughts as have colleagues at the NIA and NHLBI.

And also, my thanks to Ms. Shelly Sayre, Ms. Rachel Vojvodic, Ms. Judy Bettencourt, and Ms. Michelle Cohen, who, for twelve years, patiently bore my blackboard scratching, conversational musings, and energetic remonstrations about the current use of biostatistics in health care research.

And of course, I am indebted to giants in the field of mathematical analysis, e.g., Georg Cantor, Bernhard Riemann, Henri Lebesgue, Johan Stieltjes, and Andrey Kolmogorov, whose works were omnipresent and always ready for my study and absorption. Short biographical sketches of each are included at the end of this book.

Lem Moyé
Chandler, Arizona
February 2020

The Ten-Thousand-Foot View

Before we get to the granular details, let's get an overall perspective on my approach: the ten-thousand-foot point of view.

I believe that clinical investigators are simply interested in answering general questions authoritatively. One of the most important of these questions is "Are subjects who have received the test intervention better off than those in the control group?" In order to answer this question, we need to identify and weigh the evidence for each of benefit and harm from the collection of all analyses performed.

I am assuming for all examples (except those of the chapter on exploratory analyses) that the researchers have conducted a well-designed, concordantly* executed, two-armed, randomized, controlled clinical trial testing and intervention versus control therapy against prospectively declared outcomes of high precision. The investigators simply want an answer to the question "Are patients in the intervention group better off than those in the control group?"

Now there are many analyses that these investigators will conduct to address this question. However, the statistical community's argument of parsimony—i.e., the only analyses that are really persuasive in a clinical trial are the analyses of the primary outcomes—implies that these other outcomes and analyses, while clinically relevant, do not determine the final result of the study.

Thus, this statistical reasoning has reduced the clinical trial endeavor from its full panoply of survival, comorbidity, quality of life, and

---

*   Executed in accordance with the trial's protocol

physiologic and biologic markers, findings to the monochromatic positive, negative, or uninformative* commonly based on a single endpoint.

For example, a clinical trial assessing the impact of a new therapy on heart failure quite justifiably will choose total mortality as its primary outcome. However, investigators also assess measures of morbidity (hospitalization for heart failures and number of days patients are not hospitalized for heart failure [hospital free days]) and exercise tolerance.

In addition, they will have quantitative assessments of heart function (left ventricular end systolic volume, left ventricular end diastolic volume, and sphericity index). And they can, in addition, include a number of proteomic† measures, such as brain natriuretic peptide. A comparison of these measures across the two therapy groups provide an assessment of the change in the subject that may have been produced by their therapy.

Yet these outcome measures beyond the single primary outcome are only considered in a secondary or auxiliary role in the standard clinical trial analyses. While it is not fair to say that they are ignored, they are de-emphasized. Why that is so will be discussed in chapter 2.

Standard statistical treatments do not permit the quantitative combination of different analysis results in a clinical trial into a single expression of effect. This is not a failure of statistical hypothesis testing as much as it is the inertia in a field that analyses and interprets one outcome at a time.

Yet in health care, physicians must analyze multiple findings simultaneously; we must integrate them. This integration is conducted cerebrally, not mathematically, in clinical practice and has historically followed the same development in clinical research. In this book, we will produce an ensemble summary of the results of all analyses executed in a clinical trial that are responsive to a particular question.

Thus, the two methodologic goals of this book are to, from the entire set of analyses conducted in a clinical trial that are carried out to address a particular question, (1) examine the finding of each analysis, parsing out the component of the finding that supports benefit; (2) channel these results into and through a benefit function; and finally (3) accumulate these benefit findings over all the analyses (accumulation is principally the same as *integration* in mathematics). We will also carry out the same

---

* An uninformative result is a finding that is not statistically significant, but underpowered.

† The identification of proteins that are produced from specific organs whose presence can indicate the degree of health

procedure for each analysis, extracting the components of analyses that support harm and using duality theory.

What Is Duality?

Duality is the property of an estimator in clinical trial that allows it to simultaneously provide evidence of benefit and a finding of harm. While this property can be confusing to traditional statisticians, it is nothing new to clinicians who become accustomed to handling lab tests where findings are unclear.

As an example, consider a physician who orders a baseline serum, creatinine level, on a patient who is about to start a course of a nephrotoxic drug. This baseline finding is 0.95 mg/dl.

Completely normal.

The physician then has the measurement repeated after the patient has been on the medication for several days.

The repeat value is 1.08. The upper limit of normal is 1.1.

While it is possible for the physician to decide that the creatinine level is still normal, many doctors would give this second estimate some additional consideration. Undoubtedly, the creatinine value is still "normal" and supports the notion that the drug has not been harmful.

However, the single value of 1.08 permits another perspective. There is the variability introduced by the (im)precision of the estimate. In addition, there are physiologic effects that could change the creatinine value, e.g., the patient's hydration state, and of course, the drug itself. The value of 1.08 might reflect the beginning of toxicity.

Put another way, there is a region around the value of 1.08, which we might call the region of plausible creatinine levels. Part of the interval may reflect normal creatinine values; another part of the interval (that which is greater than 1.1) may reflect abnormal values. The single value of 1.08, because of its imprecision and the myriad influences on it, generates a wide plausibility interval and simultaneously supports normality and abnormality.*

---

* Of course, the initial value of 0.95 also has a plausible range of values. However, one can incorporate this plausible range by identifying the plausible interval for the difference between the baseline and follow-up creatinine levels without loss of generality. This is not carried out in this example to simplify the presentation.

This is what is meant by duality; a single estimate can reflect the possibility of benefit (or in this example, no harm) and the possibility of harm.

As another example, consider the current debate about the role of peanut oral immunotherapy. A response to this allergy is the use of peanut immunotherapy, where the subject with the allergy is gradually given an increasing dose of peanut paste over a course of weeks to decrease their immunosensitivity to the legume.

A recent meta-analysis[*] of clinical trials that each examined the role of immunotherapy demonstrated that subjects in the immunotherapy treatment arms had a greater rate of surviving an oral challenge upon concluding treatment than the control group (relative risk of 12.42 [95% confidence interval of 6.82–22.61]. At first blush, this appeared to be a success; however, the authors also noted that patients in the immunotherapy groups of these trials had a greater incidence of anaphylaxis, anaphylaxis frequency, and epinephrine use. How could both findings be true?

An examination of the description of the results by the authors and also a commentary[1] revealed the answer. Many individuals in the treatment group were able to complete the exposure program successfully. However, in that same group, individuals fluctuated in their reaction to the peanut paste, sometimes reacting to a dose to which they evinced no allergic reaction previously. The responses to the therapy were not just variable but were complex, demonstrating a pathophysiologic intricacy that undermined the contribution of the standard statistical estimator to a helpful understanding of the exposure's effect.

The same therapy produces harm in some individuals and benefit in others. This is the essence of duality. In duality, the estimate of effect reflects a range of values, some consistent with benefit and others consistent with harm.

---

[*] Meta-analyses that combine studies not designed to be combined, however mathematically elegant, can be briar pits when it comes to interpretation. The purpose of referencing one here is simply to provide an example of duality, not to provide a dialect for or against this methodologic approach.

Plausible versus Confidence Intervals

Many readers will recognize the similarity between a plausible interval and a confidence interval. Both are intervals around a statistical estimator (e.g., a sample mean difference) that reflects variability of that estimate. However, this is really the only similarity.

Confidence intervals were developed to reflect the sample-to-sample variability of the estimator. That is the only variability that they were designed to capture. Plausible intervals capture that variability but, in addition, also commandeer other sources of uncertainty. For example, consider the technician-to-technician variability in the assessment of an MR image. This is not sampling variability; it is the imprecision in the use of the measurement tool itself.[*]

Secondly, plausible intervals have no formal estimate of confidence. They are not 90% plausible or 95% plausible. That has no meaning for us here. It is simply a region of values that are believed to be credible based on the inaccuracy (imprecision and sample-to-sample variability) of the estimator and any bias introduced by the research design and execution. Plausible intervals are, in general, wider than 95% confidence intervals.

Finally, plausibility regions need not be symmetric.

This Book's Methodology—Parse, Channel, and Accumulate

This book develops three processes and compares the result.

The first process is to parse the plausible interval into one interval that suggests benefit and another suggesting a harmful effect.

Next, we will channel that benefit interval into and through a benefit function. We will repeat this process for every analysis that the investigators believe is responsive to the question "Are patients in the intervention better off than in the control group?"

Finally, we will accumulate these unitless measures through integration. We then carry out the same process for all the plausible intervals of harm and then compare the two.

---

[*] Precision is the ability of different measures on the same subject at the same time to be as close to each other as possible. Variability is the difference in the measurement across different subjects in different samples.

Set Theory and Quanta Analysis

However, this attempted accumulation of estimates of benefit and estimates of harm raises several critical questions.

The first is that many of the analyses in a clinical trial use the same observations and the same variables; they are redundant. Shouldn't subsequent analyses, using many of the same observations and variables as previous analyses, be discounted? After all, those observations and variables (in the guise of other estimators) were already used.

Each analysis is based on a combination of subjects and variables. We need to keep track of not just the raw number of them but also their actual identities in order for us to follow the redundancy.*

We call this data, this collection of observations and variables used for an analysis, that analysis's *region of the analysis*. We need to compute the size of this region and track its overlap with the region of other analyses.

Set and measure theory permit us to measure the size of this region. We will call this size its ψ–measure (pronounced as "psi measure").

Different analyses will have different regions of analyses (since they use different collections of observations and variables) and therefore different ψ–measures.

When the regions are disparate (that is, the collection of analyses use entirely different subjects and variables from one another), their ψ–measures add. However, when these regions have overlapping subjects, the ψ–measure has to be computed differently.

Measure theory suggests that analyses be broken into analysis fragments or quanta, which reflect contributions to the overall ψ–measure that are independent from other quanta. We use set and measure theory to compute this accumulation of ψ–measure over different but intersecting regions of analysis.

This accumulation is the total content (denoted as Γ_q) of the analyses used to address question q. We simply write this as $\Gamma_q = \int_{A_q} d\psi$.(figure 1).

---

* Subjects must, of course, be de-identified to meet with HIPAA rules. Identification here means simply their study ID number or acronym.

Weighing the Evidence

$$\int_{A_q} d\psi$$

Integral sign announces Our intent to accumulate → ∫

Accumulation is according to the rules of ψ - measure

Accumulation is over the region of analysis that addresses question q

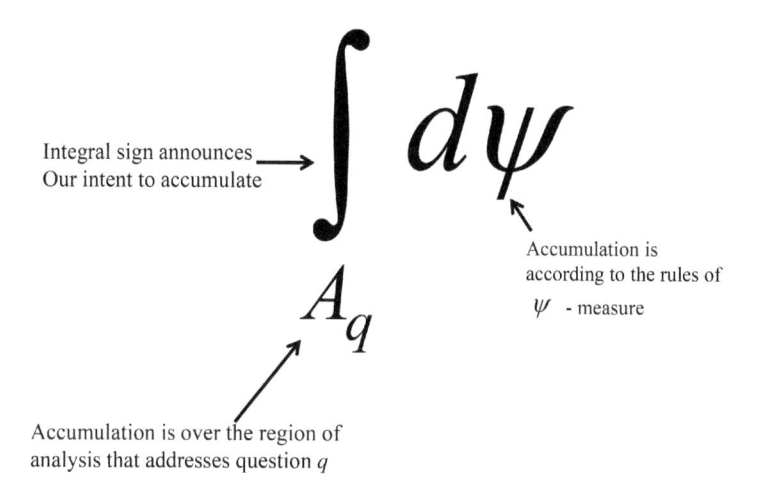

Figure 1. How to the integral over of set of clinical research analyses.

Those readers who do not have strong backgrounds in mathematics should not be frightened of this notation. Figure 1's integral is nothing more than an announcement of intent. It states that we intend to accumulate all the analysis content over regions of A_q, where analyses were conducted to answer question q. These regions have their content assessed using ψ–measure.[*]

What is unique for us in the clinical trial arena is that here we are integrating over not just part of the real line (like in a first calculus course), but over a set of analyses. This concept is not novel in mathematics, but it typically is not applied to clinical research. This type of integral is special in measure theory going by the moniker Lebesque-Stieltjes[†].

We can now accumulate the benefit and harm functions from duality theory and accumulate them with respect to ψ–measure to obtain ensemble measures of benefit and harm. This is what duality-quanta analysis attempts to accomplish.

If we were going to say this mathematically, we might begin by identifying a collection of analyses $\{\omega_1, \omega_2, \omega_3, ..., \omega_n\}$, which

---

[*] Note that this integral is not our classic one, where we integrate over a region of the real line (e.g., the area under the Gaussian or normal curve) and use familiar assessments of area, such as dx or dz.

[†] Pronounced as "LeBĀk-StillJes"

are to be analyzed sequentially. For each of these analyses, say, the ith analysis, we compute the plausible interval for benefit, $\chi_i^{(b)}$, and apply the benefit function to it, $\mathbf{Y}_b\left(\chi_i^{(b)}\right)$. We then accumulate or measure the benefit function over each of the n analyses using its contribution $\psi(\omega)$. In mathematics we would describe this as accumulating the benefit function over all regions of analyses that are included with respect to the ψ–measure, writing it as $\int_{A_q} \mathbf{Y}_b\left(\chi_i^{(b)}\right) d\psi$.

We do the same thing for harm, $\int_{A_q} \mathbf{Y}_h\left(\chi_i^{(h)}\right) d\psi$, and then take their ratio. The construction of this process and managing its complexities and implications is the main topic of this book.

So we have two concepts to balance. The first is the regions of analyses that we must mathematically dissect and manage. The second is the parsing of the plausible intervals into those portions reflecting benefit—collecting them, "measuring them," and accumulating them over the analysis region quanta—and then repeating this process for harm. Finally, we take the ratio of the two.

A second issue involves the relationships between variables. This issue of correlation is easily addressed once we have developed the notation for this quantum approach and is addressed in chapter 22.

But before we dive into those details, let's first address the question "Why is this development even necessary?"

References

[1] Abbasi, J. "Weighing the Risks and Rewards of Peanut Oral Immunotherapy." *JAMA* (July 31, 2019). Epub ahead of print. https://doi.org/10.1001/jama.2019.9142.

Statistical Theology

Choosing to walk away from an established guide in health care like the p-value—even though that guide is now quite blind—is difficult. P-values, as part of a new architecture that included prospective protocols, served well as part of an organizing framework in the 1950s, bringing structure to inchoate clinical investigation and disorderly research findings. Like training wheels on a youngster's bike, the p-value helped keep the young clinical trial enterprise upright.

However, we have been riding for seventy years now, and these structures that kept the rudimentary clinical trial infrastructure in place are now too constraining.

Clinical research has been and remains the best hope for the solution to chronic disease, whether that hope reside in genetics, preventive maneuvers, pharmacologic therapy, or biologics. Yet, the complexity of clinical trial programs—with multiple treatment arms, interim analyses, assessments of clinical findings of mortality and morbidity, as well as examinations of promising proteomics—cannot be brought to bear with their full power and authority when forced to abide by a restrictive p-value predominance.

Specifically p-values and their attendant statistical hypothesis testing do not permit the full deployment of results produced by the research enterprise. In fact, any tool that requires rigid allegiance—even though it itself is ambiguous and defies a clear definition upon which clinical investigators, epidemiologists, and biostatisticians can agree—has more of the feel of theology than of science.

Those of you who have heard all the arguments about the problems with p-values are encouraged to skip on to the next chapter. There will be nothing new for you here. However, others of you who have accepted without question that *p*-values were important, useful, and necessary (perhaps because you were told that they were so) may be illuminated by the following dialectic.

The Two-Minute Problem

Much of the world's population does not like mathematics. Finding problems in mathematics uninteresting, irrelevant, and a waste of time, most people are all too happy to turn over "the math" to someone else—whether that math be simple accounting, working through some plain geometry for determining how many gallons of paint are required to double coat a wall, determining how long it takes to get to a destination at their current speed, or figuring out a tip for their restaurant server.

However, when they are pressed to solve a math problem, the math problem falls into two different categories. The first category is the class of math problems that they can solve instantly. For example, city and state taxes combined add 10% to the initial cost of the car. The initial cost of the car is $26,000. Then the additional tax is $2,600.

Easy as pie.

All other problems fall into a second category: the timeless set. For these problems, it doesn't matter whether the individual is given two seconds, two minutes, or two years to work out the answer. They don't know how to approach the problem, much less solve it.

A major difference between these people and the mathematician is that the mathematician has tools that she can use to help convert the two-year problem into a two-minute problem.

One of these tools is simplification.

In Polya's great text of mathematical guidance *How to Solve It*,[1] an important tool available to mathematicians faced with a hard problem to solve is simply ... don't.

Instead, solve a related problem.

Commonly, that related problem can be a simpler problem. The initial mathematical problem that confronts us is complicated. Maybe it is finding the volume of water in a pool that is not rectangular, but instead has different sculptured shapes. Here, a first approach is to assume away the complications and turn the pool into a simple hemisphere. Solve the simple problem, then work back to the more complex, original one.

Or sometimes, the simplification is deemed to be sufficient. This is the issue with *p*-values.

The Cutting Room Floor

The application of statistical hypothesis testing in health care is the process by which a complex, health-related question with multiple components has been trimmed, reduced, and distilled until it produces a simple question that can be addressed by statistical hypothesis testing.

The result is commonly a simple assessment of one or a small number of clinical outcomes, leaving much of the research data and results in their richness and complexity behind "on the cutting room floor."

While this was likely not the intent of biostatisticians or the senior clinical research leaders of the 1950s, it most certainly was not the intent of physician-scientists who, in fact, collected a wealth of data in order to harvest its findings. However, the collision between the bountiful products of clinical research on the one hand and the need by administrators to evaporate this product down to a fine, alpha error–managed distillate on the other has produced a product that, seventy years later, has us scratching our heads.

This chapter discusses how we got here. Much of the following is taken from Lee Kennedy-Shaffer's fine article "When the Alpha Is the Omega: *P*-Values, 'Substantial Evidence,' and the 0.05 Standard at the FDA"[2], as well as from *Statistical Reasoning in Medicine: the Intuitive P-Value Primer*[3] and *Multiple Analyses in Clinical Trials: Fundamentals for Investigators*[4]. There is no villain in this story. We are all complicit.

Two Trajectories

From our 2020 perspective, clinical research and biostatistics are intricately intertwined. A physician-scientist would not consider conducting publishable health care research without at least contacting a biostatistician, and biostatistics, for its part, has devoted itself to new methodologies that are commonly related to applications in health care research.

However, this has only been the case since the early 1950s. Prior to that, health care research and epidemiology on the one hand and probability and statistics on the other hand, in the main, followed very different paths. Others and I have discussed this topic before. The goal here is not to recapitulate in detail but to provide the vibrancy that each field had developed in order to understand their ultimate calamitous collision, a detonation that has produced our current state of affairs.

This collision was not about the *p*-value. Like the city of Gettysburg, the *p*-value just happened to be in the wrong place at the wrong time. The battle was—and continues to be—which perspective, statistical or clinical, governs the conclusions of health care research.

Epidemiology and Health Care Delivery

Epidemiology represents a disciplined, cerebral approach to drawing health care conclusions from facts that must be discerned within a fog of variability.

For thousands of years now, these reasoning men and women focused on applying what they observed and deduced to the sick in their care. They suffered from crude instruments, absent labs, and no way to collaborate between villages, towns, and cities except through what we might describe as a verbal, then later written, crude case report.

Celsus stated that "Careful men noted what generally answered the better, and then began the same for their patients" (circa AD 25).[5] For the next 1,900 years, advances in clinical medicine occurred through the combined use of careful observations, clear recorded descriptions, and deductive reasoning. Chance observation tested and sometimes overturned standard dogma, e.g., the belief that musket wounds must be permitted to fester to heal.[6]

John Graunt established the application of deductive reasoning to multiple data points[7] and along with William Petty, developed the life table methodology, permitting for the first time the computation of number of deaths from the bubonic plague, consumption, and phthisis (tuberculosis), which could be quantified and followed over time. James Johnson[8] pointed out the value of literature review, the role of confounders, using replicates of treatment to address result variability and study replication. Applying these principles, James Lind on the HMS *Salisbury* in 1747 worked to defeat scurvy on the high seas, and later John Snow discerned the cause of cholera in London (1830–1850).

Principles of causation were elaborated by Sir Austin Bradford Hill, the father of clinical trials[9], providing tenets that were based on a commonsense approach to determining causality and are remarkably free from complicated mathematical arguments.

The critical point here is that much of this work was not mathematics-centric. For over nineteen centuries, there was little quantitative development in which they could rely, so these thoughtful men and women developed solid, intelligent, but essentially nonmathematical contributions. They would use data sets when it was available (e.g., the work of Graunt and Petty), but they had developed skills to operate in its absence.

Probability, Statistics, and Prediction

Probability and statistics developed from a different foundation where its cornerstone (the random process) was condemned as a capital offense by religious leaders during the Middle Ages.* However, decade by decade and then century by century, life softened outside the monastery walls. Villages, towns, and then cities flourished again. The advent of the Renaissance heralded the notion of free thought, and the Industrial Revolution introduced the concept of leisure time.† With leisure

---

* It is easy here in the twenty-first century to be critical of religious dicta, but in this case, they served as a protection, albeit extreme reaction, to a lawless and demented time that we know as the Dark Ages. The destruction of the Roman Empire inaugurated an unparalleled error of depravity. With Rome and other cities demolished, the only choice individuals had was to survive in the countryside where people were never safe and fledging crops planted by outcast city dwellers mostly always failed. Life for these unfortunates came down to a decision: either join the rampaging gangs or join the church. Many flocked to the monasteries not to be devout but to simply survive.

Inside these protective walls, monks and nuns outlawed gambling and its random event foundation for two reasons. First, gambling was commonly conducted for financial gain, a worldly concern more fitting for those who chose lives outside the insulating walls of the monasteries and convents. Secondly, the very concept of randomness suggested that events occurred outside the control of God. While gambling was simply a crime punished by expulsion, contemplating randomness was blasphemy.

† Up until the off-loading of manual work to machines, a person's day was consumed principally by work, time at the church, or eating and sleeping.

time, there was a new (now legal) interest in what became the raging pastime—gambling.

Gifted observers began to use the data from these activities. A major advance was produced in the early 1600s by Abraham de Moivre, who developed the theory of the normal distribution as an approximation to the binomial distribution. Fermat also wrote extensively on gambling, which he correctly perceived as a process by which the future behavior (of the game) is predicted from past experience. This was the beginning of modern probability thought.

However, the field did not escape criticism. Early tabulators involved in using techniques such as sampling in census counts were said to not be involved in science but in *political arithmetic*, defined as "the art of reasoning by figures upon things related to government."[*]

As the work of Laplace, Poisson, and others moved probability forward and delved into the implications of the work of Thomas Bayes, an interesting debate arose about the contribution of the field to society: should the quantitative scientists be the best ones to interpret the data that they analyze, or should that be turned over to another?

In 1834, when the Statistical Society of London (later to become the Royal Statistical Society) was formed, they lent their perspective to the debate through their selection of an emblem: a fat, neatly bound sheaf of healthy wheat that represented the abundant data, neatly collected and tabulated. On the binding ribbon was the society's motto: *Aliis exterendum*, which means "Let others thrash it out"[10]. As this sense of the field took hold, Pearson and Gossett pushed the work of what to do with aggregate data up to the brink of statistical inference to the twentieth century.

Eruption

The two fields of epidemiology and statistics did more than peacefully coexist; they, in fact, worked jointly on major issues, such as demonstrating that early vaccines for smallpox were effective.[†]

---

[*] From Charles D'Avenant, taken from Karl Pearson's *The History of Statistics in the 17th and 18th Centuries*

[†] One of the first examples of comparing observed results to what was expected. Here, the Bernoulli brothers, using the binomial probability distribution, computed the expected number of cases, and the epidemiologists and health

However, a tempest was coming for them all in the twentieth century, and it was the clinical or observational scientists and epidemiologists who were first blown off trajectory of progress. Yet this storm's creator was not Ronald Fisher and his notions of significance testing in the 1920s.

It was Albert Einstein.

While not trained in biology, medicine, or the observational work of epidemiologists, Einstein created the intellectual, seismic disturbance that shoved the observationalists' ships perilously close to the rocks.

In deriving the famous equation $E = mc^2$, Einstein developed the principle that positions of reference were relative. Observers from different platforms could observe the same result and come to different conclusions, and both were right.

Geologists, biologists, chemists, clinicians, and epidemiologists had struggled over the centuries to develop experimental paradigms that would provide the best and unbiased platform from which to observe a result and therefore draw a correct conclusion. Einstein told them that this was impossible. By showing observational scientists that what they observed may not be as close to the truth as their training suggested, Einstein, unwittingly but firmly, invalidated them.

Some chose to challenge him.

The book *One Hundred Authors against Einstein*[11] was an attempt by nonphysicists to resist the great physicist's ideas of time dilation, declaring them irrelevant to biologic processes. It countered that little could be learned of the real world by abstract mathematics, which itself had lost its connection to common sense.

Yet the proof of the general theory of relativity removed most major scientific criticism of Einstein's work. Einstein's second compendium on gravity demonstrated to observational scientists not only that they could not trust their instruments but also that mathematics was more trustworthy. In demonstrating that an observation (the bending of light by the sun) had been missed through thousands of years of sol observations but had been predicted by mathematics was not just a tour de force of physics but also an insufferable blow to those who believed in the power of observation.

The only encouragement the observationalists received from the new masters of physics was to not trust their eyes but to instead rest their faith on mathematics. And this mathematics was new, dense, and to them,

---

care providers counted the observed, permitting a comparison between the two.

impenetrable. How were they supposed to, for example, listen to hearts or interpret chemistry tests at relativistic speeds, and how could that possibly help them in their work or practice?

They were dead in the water.

Meanwhile, statisticians, who themselves were struggling with the concept of how to use data to convincingly answer questions, received an unanticipated new wind in their sails from Ronald Fisher.

The 1920s and the Appearance of Statistical Significance

One of Ronald Fisher's earliest writing on the general strategy in field experimentation was his 1925 book, *Statistical Methods for Research Workers*[12], and in a short 1926 paper entitled "The Arrangement of Field Experiments"[13]. This work contained Fisher's thoughts on experimental design and his initial framework for significance testing.

It is also where the first mention of a 5% level of significance first appeared.

Using as an example the assessment of manure's influence on crop yield, he puzzled over how to compare the yields of two neighboring acres of land, one treated with manure and the other not. It was true that the manure-treated plot produced a 10% greater crop yield than that of the nontreated plot, yet Fisher knew that there was also variability due to other factors (e.g., soil moisture, insect density, difference in seed quality, etc.). Fisher distilled the question down to an assessment of how likely would one expect to see a 10% increase in crop yield in the absence of the manure by chance alone. He then reasoned:

> [T]he evidence would have reached a point which may be called the verge of significance; for it is convenient to draw the line at about the level at which we can say "Either there is something in the treatment or a coincidence has occurred such as does not occur more than once in twenty trials." This level, which we may call the 5 per cent level point, would be indicated, though very roughly, by the greatest chance deviation observed in twenty successive trials.[14]

He added,

> If one in twenty does not seem high enough odds, we may, if we prefer it, draw the line at one in fifty (the 2 per cent point) or one in a hundred (the 1 per cent point). Personally, the writer prefers to set the low standard of significance at the 5 per cent point, and ignore entirely all results which fail to reach this level.[15]

Fisher continued to say that if he had the actual yields from earlier years and could compute the variability of the yields, then he might use Student's *t*-tables to compute the 5% significance level.

This was the birth of the 0.05 level of significance.[16]

Observational Scientists React

Many statisticians were elated by Fisher's writings. The notion of sample-to-sample variability had been known to investigators for years. Early twentieth-century Bayesian and non-Bayesian statisticians alike wrestled with how to combine the two in order to assess the value of a measured, experimental effect size. Finally, here was a way to deal with the chronic problem of sample-based research using objective mathematics

Yet Fisher's point of view on experimental design became the flash point of a new controversy. Many observationalists, believers in the scientific method, welcomed an approach to test a scientific hypothesis. Yet to them, the significance testing scenario was counterintuitive, representing the unhelpful type of thinking that was likely to be produced by mathematical workers who did not spend sufficient time in the observationalists' world of data collection and deductive reasoning.

To these observationalists, the entire process of statistical hypothesis testing was reversed. The scientific method began with the notion of a hypothesis that the scientist believed, e.g., a new, compounded powder will reduce fever in postpartum women. This was then to be supported or disproven by the collected data.

However, statistical hypothesis testing began with the reverse perspective that the compound would not be effective. It then set up the assessment so that the collected data would disprove (i.e., nullify)

that hypothesis. This not only was ungainly and complicated but it was also indirect.* Specifically, this new, upside-down paradigm of statistical significance appeared to deny the scientist the ability to prove the hypothesis he believed was correct. Instead, the scientist would be required to replace the strong assertion of his own affirmative scientific hypothesis with the tepid alternative of disproving a hypothesis that he did not believe.

To traditional observationists, Fisher's significance testing appeared to be just the type of indecipherable, mathematical, reverse logic that had already shaken the foundations of early twentieth-century epidemiology. Already bruised by the two decade-old assault on its philosophical opinions by physicists and mathematical theorists†, they gathered their forces. From the epidemiologists' point of view, it was bad enough that they had to sit still where Einstein's advocates criticized their world of

---

* Statistical hypothesis testing had very much the look and feel of proving that the $\sqrt{2}$ was irrational. One did not actually show that this quantity was an irrational number. Instead, one assumed that it was rational and then reasoned to a contradiction. Note that there is no special feature of an irrational number that is revealed in this proof, only that it is not rational.

† Over time, epidemiologists have successfully defended their time-tested methodologic perspective. Of course, the flaw in all of the criticisms regarding the use of observation as a foundation method of epidemiology lies in the difficulty in translating findings that are germane in one field (physics) to that of another (life sciences). While the findings of the relativity laws are, in general, true, they are most useful in physics. The theoretical physicist may be correct in asserting that every observer is biased and that there is no absolute truth about the nature and magnitude of the risk factor–disease relationship. However, this does not imply that all platforms are equally biased. Epidemiologists never stopped striving to find the most objective position possible. Certainly, if bias cannot be removed, it should be minimized. The fact that bias may not be excluded completely does not excuse its unnecessary inclusion.

Second, while mathematicians are capable of predicting results in physics, they have not been able to predict disease in any important or useful fashion. No mathematical models warned obstetricians or their pregnant patients of the impending thalidomide–birth defect link. Similarly, mathematical models did not predict the birth defects that mercury poisoning produced in Japan. While physics often studies processes in which mathematics can reign supreme, real life and its disease processes have proven to be painful, messy, and chaotic affairs. The substantial role of epidemiology is incontrovertible in the development of the most important, new health care research tool of the twentieth century—the clinical trial. The time-tested tools of epidemiology continue to prove their utility in the present day.

observational scientists. However, they did not have to take this from an unknown agrarian statistician.

The field responded vehemently and vituperatively, as in, "What used to be called judgment is now called prejudice, and what used to be called prejudice is now called a null hypothesis … it is dangerous nonsense …"[17]

In the meantime, enthusiasts in the statistics community for the notion of significance testing grew. In the 1930s, Egon Pearson and Jerzy Neyman developed the formal theory of testing statistical hypotheses even further. In addition, they introduced the notion of statistical power[18], while other workers produced the concept of the confidence interval[19, 20, 21]. These developments took place on the searing bedrock of controversy, fueled by the vitriolic criticisms of Berkson[22, 23] and vibrant ripostes by Fisher[24].

Enter the Administrators

Both sides of this debate were affected by the explosion in health care research in the 1940's. World War II—with its requirement for new, improved medicine and delivery of health care services to both soldiers and refugee populations—generated explosive new waves of health care research.

Medical groups in the UK and the US developed and tested new medications, e.g., antibiotics and oral hypoglycemic agents. The Medical Research Council (MRC) was conducting one of the first clinical trials to evaluate streptomycin. New investigators were pouring into the labs of pharmaceutical companies or the facilities of universities.

Few knew anything about the raging controversy concerning significance testing. However, new, statistics-centered thought was beginning its injection into clinical research.

W. Edwards Deming, in his 1943 book *Statistical Adjustment of Data,* suggested using *p*-values in repeated experiments as measures of the quantum of evidence against the null hypothesis. Specifically, he recommended the use of "statistical significance" as an inferential method. However, he also added the monitory that "[s]tatistical 'significance'" by itself is not a rational basis for action."[25]

This warning was not headed by leading journals.

A 1950 editorial in the *Journal of the American Medical Association* (JAMA) under the banner question "Are Statistics Necessary?" offered a wholehearted answer: yes. According to the journal, investigators developing new therapy must be able to (1) assemble a table comparing treated and control subjects and (2) be proficient in computing *p*-values.[26]

Another JAMA article required statistical tests to be the basis of evidence for therapeutics, urging clinical studies to use randomization, untreated controls, and significance testing.[27] The *Annals of the New York Academy of Sciences* similarly called for quantification of clinical trial results and the application of statistical reasoning.[28]

Reaction to this building momentum for statistical significance was pointed, with much of it negative. In 1960, William Rozeboom wrote of his concern for the use of statistical hypothesis testing. Not only did he resist the notion of accepting or rejecting a scientific hypothesis completely but he also resisted the notion of the 0.05 significance level, recognizing that there was no scientific underpinning for the 0.05 level.[29]

Presciently, in a speech delivered to the International Biometric Society in 1969, the outgoing British regional president J. G. Skellam warned that significance tests might "exercise their own unintentional brand of tyranny over other ways of thinking."[30]

However, these iconoclastic perspectives were overwhelmed by the medical research literature, which was in a state of tortured turmoil over how to manage data assessments. While there was a time-tested format for reporting case reports (journals had, after all, been publishing these for generations), there was no standard for reporting—much less analyzing—a data set.

All agreed that having more than one observation improved upon a simple case report. It remained an open question as to how to report this data. "What are the measures of central tendency? How should dispersion be reported and managed? And how does one compare different experiences when those experiences are segregated by treatment group and control group?" These questions plagued journal editors who were anxious to find an organized and controlling structure.

However, a move to statistical hypothesis testing played to the journals' best practical interests as well. At the time, the number of manuscript submissions were swelling with no crest in sight. The introduction of a significance testing requisite would function as a constraining factor, modulating the number of articles worthy of submission at a time when journal editors were overwhelmed.

Rule of Thumb

The US Food and Drug Administration (FDA) now entered this tempestuous arena as they themselves struggled with the absence of analytic rigor in new drug applications, the number of which exploded during the postwar era.

From the 1950s into the 1960s, the state of protocol submission to the agency was abysmal. Many drug applications were submitted to the FDA, not just without a statistical plan but also without a protocol.[31] The agency soon joined the chorus of voices calling for statistical rigor.[32]

There was no statement in the regulations that the 0.05 p-value metric had to be used, but there was an understanding that the customary rule of thumb in assessing the effect of an experiment would be the Fisherian 0.05 standard.

Thus, by the midtwentieth century, overwhelmed journal editors as well as overworked FDA employees received a welcome architecture to both provide research-supporting structure and also serve as a bulwark and filter to against the applications they were receiving. Sometimes tacit and sometimes explicit, there was an understanding that statistical hypothesis testing would be required for drug application submissions at the FDA and for manuscript submissions to journals.

From a practicality perspective, one cannot blame these administrators for seizing upon statistical hypothesis testing. The field was overwhelmed with poorly designed research and shoddy analysis plans. Introduction of a new methodology promised much needed injection of rigor.

The p-value appeared to be the natural solution. After all, it incorporated the size of the sample, the effect size that was observed in the study as well as its variability, plus an attempt to assess the "generalizability" of the results. And of course, it was simple to interpret. Values < 0.05 were considered acceptable. Those > 0.05 were not worthy of further consideration. It was certainly compact and believed to be quickly interpretable.

A more thorough discussion of this complex choice is available.[33, 34, 35, 36, 37] The installation of the p-value as a research result metric was not one of malevolence. Instead, the decision to move forward was made by observant and overwhelmed administrators who simply wished to practically underwrite and promulgate the most solid research efforts. They hoped that the use of this tool would permit the data to speak for themselves in a structured fashion, free of the bias that an investigator would bring to the research paradigm.

Nevertheless, these decisions by these influential gatekeepers of health care research had a profound stifling influence on not just the structure of research reports but also on the climate of research itself.

Transmogrification

In the 1960s, as the FDA was incorporating significance testing into its new drug, epidemiology solidified its antithetical approach to the role of hypothesis testing in medicine through the elaboration by Sir Austin Bradford Hill—first in 1953[38], then again in 1965 of the well-established epidemiologic tenets of causality. These tenets served as the basis of epidemiologic causal thinking in the midtwentieth century.*

However, Hill's thoughtful, accepted approach was beginning to be supplanted by the following style of reasoning: "Since the study found a statistically significant relative risk ... the causal relationship was considered established."[39]

While this type of comment was not typical, it did demonstrate the extreme conclusions that were beginning to be based solely on the p-value.

The answer to the central question "What is the role of mathematics in drawing conclusions from health care research?" now seemed to be that mathematics was going to play the predominant role, with little need for additional thought.

What had been offered by Fisher, Neyman, and Pearson, as an objective sense of the strength of evidence of research result, was now being transmogrified into a popular but inadequate substitute for clear, causal thinking as workers replaced their own careful, critical review of a research effort with the p-value. Was there to be no further role for assessing clinical significance in the absence of statistical significance?†

---

* Free of complicated mathematics, these hallmarks of a causal relationship have twin bases in common sense and disciplined observation. The nine precise Bradford Hill criteria are (1) strength of association, (2) temporality, (3) dose-response relationship, (4) biologic plausibility, (5) consistency, (6) coherency, (7) specificity, (8) experimentation, and (9) analogy. These are well elaborated in the literature.

† One of my medical school interviews took a disastrous turn for the worse when

Also, more nefariously, scientists began to sculpt and therefore promulgate their findings based on analyses in which the *p*-value was small.*

Pushback

It was inevitable that some workers in health care, as their forbearers had fifty years earlier, vigorously resisted this degradation in the scientific thought process. The dispute broke out into the open in 1987 when the prestigious and well-respected *American Journal of Public Health* solicited an editorial arguing that significance testing be purged from articles submitted for review and publication.

Subsequently, the epidemiologist Alexander Walker debated with the statistician T. W. Fleiss over the use of significance testing[40, 41, 42, 43, 44] which was joined by Poole[45]. In addition, Rothman, in an editorial for *Annals of Internal Medicine* in 1986, wrote that "[t]esting for statistical significance continues today not on its merits as a methodological tool but on the momentum of tradition."[46]

Doubling Down on the P-value

In the meantime, cardiology began to get a taste of what was coming from its forced feeding of *p*-value laden meals.

- The suggestion by the Multiple Risk Factor Intervention Trial (MRFIT) that hypertensive men with baseline ECG abnormalities were harmed and not helped by antihypertensive therapy was a stunning blow to the hypertensive control community, who were just as stunned to later realize that this

---

the interviewer, during his examination of me, received a letter of rejection of his submitted article from a journal because the results did not reach statistical significance. The investigator-interviewer raged at the inappropriate use of statistics, which was, after all, meant to describe and not to decide research matters. By the time the interviewer soothed himself, the interview time ended, and I was sent on my way.

* This now goes by the sobriquet "*p*-hacking."

clinical trial–based result was false, based on a *p*-value, and was not reproducible.[47]
- The clinical trial–based finding that the anti-acetylcholinesterase therapy vesnarinone could save the lives of patients with heart failure was reversed by subsequent clinical trials that demonstrated the harm of this therapy.[48, 49]
- A clinical trial demonstrating the mortality benefit of losartan had its result overturned by a subsequent clinical trial.[50, 51]
- A clinical trial–based subgroup analysis that declared amlodipine could prolong the life of patients with nonischemic cardiomyopathy was also reversed by a subsequent clinical trial.[52, 53]
- The diminished and confusing efficacy findings from United Kingdom Prospective Diabetes Study (UKPDS)[54, 55, 56, 57, 58]
- The violation of prospectively declared analysis procedures in the Lipid Research Clinics (LRC) trial[59, 60]
- The US carvedilol program controversy[61, 62, 63, 64, 65, 66, 67] of the late 1990s*

This collection of results served to undermine the confidence of cardiologists in the interpretation of clinical trial results. Minna Antrim could have been speaking to them when she said in 1901, "Experience is a good teacher, but she sends in terrific bills …"[68]

At this critical juncture, clinical trials in cardiology were becoming more complex, with multiple prospectively declared endpoints, multiple treatment arms, subgroup evaluations, analyses, and the very early examination of proteomics results. The diseases that were studied were complicated, and investigators—following their natures—wanted to be sure that they captured as much of the dimension of the offered as they could.

If the *p*-value wasn't serving its anticipated role, then other quantitative research tools were required. Yet, the response of the biostatistical community was to double down on the *p*-value, producing the following two research principles.

The first was to assert the dominance of prospectively declared research efforts. This was in effect a sustainment of the priority of the research's protocol; it was the rule book of the research effort.

---

* The author played a role in this controversy.

The second was that endpoint assessments must be planned with clear and declared assessments of type I error penalties.

The first was a re-enunciation of the established need in the 1950s to have a protocol in place. It formalized the research thinking, stating the investigators' belief in the effect and mechanism of action of the intervention. It also stated in practical details the needs of the research, permitting the investigators to identify equipment, expert committees, and computing facilities necessary to conduct research efforts with precision and accuracy.

However, the reinforcement of type I error penalties would have profound implications for clinical research.

Clinical Research Complexities

In my view, the reanchoring of p-values in the 1990s as a requirement in the dynamic clinical-research environment with its new complexities was a mistake. The estimate of its beneficial effects with its introduction in the 1950s was conflated with the requirement of a structured protocol, a requisite that paid handsome research dividends.

However, the p-value did not add to the stronger research foundation, already protocol-fortified. It instead infused a confused metric, which has only served to distract investigators from the need for clear causal thinking when studying complex diseases.

By the beginning of the twenty-first century, this injection was to be stultifying.

In clinical research, research interpretation would now operate in a tightly controlled type I–error environment. Specifically, only outcomes that had type I error allocated would be considered as primary[*] as a response to a situation in which nonprotocol–tethered research actions were producing clinical trials purportedly answering the same questions, but with disparate results.

By doubling down on the p-value, biostatisticians and others were requiring that the investigators focus on a small number of outcomes. And, since statistical power decreases for smaller alpha levels, ceteris paribus, there could not be very many of these outcomes because the overall (or family-wise) type I error had to be 0.05.

---

[*] A primary outcome is one on which the trial's effect of therapy is classified as positive, null, or harm.

Most clinical trials have only one primary outcome with a much smaller number of trials having two or more primary outcomes. However, no matter how many outcomes were prospectively declared, the only outcomes that would "matter" were those that had an alpha level declared prospectively.

Thus, investigators were then forced into the position of selecting the outcomes that they thought would yield a positive result in order for the trial to be judged positive. For some experiments, it is clear what the outcome should be. There may be an expectation in the research and clinical community, or the regulators may have outlined what outcome they wish to see.

However, many research programs have an idea how the therapy may work but had no reliable guiding pre-assessment as to what outcome is going to be positive in one particular sample of patients. In one sample, it may be left ventricular ejection fraction. In another sample, left ventricular end diastolic volume may be most influenced by therapy. Each has known variability, and each is related to similar clinical consequences, yet the investigator cannot know which of these outcomes will be influenced by the therapy in this sample. Yet statistical hypothesis testing requires that they choose one or choose both and pay a severe type I error penalty.*

Physician-investigators understand that there is a wide breadth of knowledge to be gained by analyzing the entire data set. Insisting on parsimony (that is, focusing on and analyzing only the small number of outcomes that have type I error allocated for them prospectively) means that much of the data will essentially go unanalyzed.†

Investigators want to cover new ground and enjoy the exploration process. Exploratory analyses can evaluate the effect of the therapy in subgroups, the effect of the therapy on different endpoints, and the effect of different doses of the medication.

The rationale for identifying a large number of outcomes in clinical research has its own undeniable force of logic (stemming from logistical, financial, and epidemiologic rationale, as well the need to examine data

---

* They could choose both but have to divide the type I error between the two, leading to an important increase in sample size.

† In 2016, an NIH admistrator said to me that if it were up to him, the only clinical trial outcomes that would be funded would be those that were an affirmative part of the type I error allocation. While this is not representative of NIH policy, it does reveal how penetrating the type I error mentality can be.

in new and provocative ways—exploratory analysis). Complex clinical-research endeavors were expanding beyond the *p*-value's ability to guide helpful interpretations.

However, the 21st century statistical paradigm of controlling the overall type I error cut much of this off at the knees. Specifically, the community, rather than rethink the role of the *p*-value, decided to constrain research to fit into its narrow interpretative environment.[*]

The limitations of statistical hypothesis testing were clear. Yet, over time, these principles were absorbed by the cardiovascular community. Contemporaneous protocol review committees (PRCs), Data and Safety Monitoring Board (DSMB), the FDA, top-tier journals, and knowledgeable audiences of international cardiology meetings now expect these conditions to be met.

However, at what cost?

The results have been investigator frustration and tragic alpha-error fiascos, e.g., the MERIT-HF program.[†] This is the product of the process in which clinical researchers permitted statistical hypothesis testing to not merely be supportive but to dominate their efforts.

Wasteland

We had the opportunity to de-emphasize alpha errors and instead embrace the full panoply of findings in clinical trials by replacing the *p*-value and its restrictions.

Those who called for this in the 1990s could have been better supported by the National Institute of Health with the creation of a new methodologic area for the development of new tools to replace statistical hypothesis testing.[‡]

Bayes procedures, always a useful counterpoint to standard statistical hypothesis testing, could also have received new encouragement for

---

[*] I must point out that I was a part of this process. Having been trained in the construction, use, and interpretation of *p*-values, I regretfully supported this notion of type I error control.

[†] In MERIT-HF, the alpha allocation and endpoint composition was changed at several points in the study, ultimately producing a failed result.

[‡] See the conclusions for how this could have and might still be developed.

development.* Such devices when fully engaged permit research efforts to assess all the data and relevant analyses, providing a summary conclusion.

Instead, the research community acquiesced to the continued enforcement of statistical hypothesis testing implementation, forcing many, if not most, of the analyses in a clinical research effort outside that effort's interpretative paradigm.

The outcome has been dissatisfying and disappointing. Clinical trial protocols now include intense details about the order of analyses and what type I error function is implemented to control the overall alpha-error level. The permitted illumination provided by secondary endpoints is reduced,† and truly new and unanticipated information is extinguished and considered not publishable in many areas. Instead, analyses are highlighted that have limited role in understanding the pathophysiology of a disease or are the product of complex outcome combinations‡ but produce executable sample sizes, again driven by statistical hypothesis testing concerns.

The practical impact of these functions is to exclude the impact of analyses that can shed light on either the breadth of the findings or the biologic mechanism on the overall finding of the study by denying them the status of primary since the alpha calculus prohibits too many primary endpoints.

The clinical research mantra used to be "If the study wasn't published, then it might as well not have been done." The operational mantra now is "If alpha was not allocated prospectively for the analysis, the analysis is vacated and non-admissible."

This philosophy reduces major clinical trials with well-planned, multitudinous analyses to rest on the findings of a single primary endpoint when biology, physiology, and pathophysiology and common

---

* To its credit, the Medical Devices Division at the FDA has affirmatively placed an emphasis on this approach for at least the past decade.

† They are reduced because they are not given the same weight as primary analyses, and there is no standard way to combine primary and secondary analyses into an omnibus measure of effect.

‡ Combined outcomes are outcomes that have important pathophysiologic rationale in the study, but because of their low event, rates cannot stand as the study's primary endpoint because the required sample size would be prohibitive. These endpoints are therefore combined into a more complex endpoint requiring a lower sample size.

sense say the findings are best interpreted in the light of multiple endpoints.

The end result in many fields, e.g., cardiology, is that we now operate in a wasteland of barren research effort, stripped of its epidemiologic richness and relentlessly patrolled by a ruthless *p*-value–centric metric.

And it is about to get much worse.

Reproducibility

There are many concerns about the absence of reproducibility in health care research. Several examples have been provided here (*vide supra* vesnarinone, losartan, amlodipine) where the findings of one (expensive) trial were overturned by another. Examples are commonly provided about the reproducibility of research in other sciences, and many wonder what can be done to improve the reproducibility in clinical research endeavors.

In the journal *Nature Human Behavior*[69], a collection of distinguished and experienced statisticians and quantitative scientists reviewed the issue in its complexity.

Their conclusion is that the health care field needs a new *p*-value threshold. Specifically, the *p*-value threshold should be reduced from 0.05 to an 0.005 level of significance.

While some believe that this might be a helpful change when applied retroactively[70], when looking forward, the sample size and financial and logistical consequences to be borne by the clinical investigators, if not insurmountable, are considerable.

To some degree, this statistician-based initiative of reducing the *p*-value threshold to 0.005 was predictable. The initial *p*-value injection into health care research was supported by many of us in the 1950s. However, its use in the presence of the new research complexities of the 1980s and 1990s (subgroups, multiple treatment arms, and multiple endpoints) generated a collection of *p*-value conundrums, yet we were not sufficiently compelled to abandon it.

Instead, we enforced its use, ensuring that type I error was conserved, to the detriment of research design.*

---

*   Rather than focus on differences in inclusion or exclusion criteria and differences in endpoint definitions

However, this tightening of the alpha imposition in the 1990's did not solve the reproducibility issue, so statisticians now suggest a further crackdown, reducing the maximum type I error from 0.05 to 0.005.

We can understand how biostatisticians would be attracted to this. We, statisticians, have been in the business of generating p-values in health care research for almost seventy years. It is a good, consistent business for us. The notion of strengthening the p-value can be packaged to produce fine publicity optics and does not change the trappings of the underlying research enterprise.[*]

However, its clinical and research consequences, if implemented, would be crushing. Clinical research efforts would increase profoundly in sample size. They would take longer to complete. Furthermore, their financial costs would skyrocket during a time of diminished financial resources.

And of course, it would not solve the problem. The reproducibility problem would likely swell. This is because the problem with reproducibility is not the p-value.

Reproducibility is a very practical issue. There are many reasons that research efforts are not reproducible. Sampling error is only one of them, and that is what p-values measure and manage. They do not measure whether patient populations are the same. They do not assess whether exposure to the intervention is identical across studies. They do not evaluate whether follow-up time is equivalent between studies.

In addition, different parts of the country (or the world) recruit different individuals with different phenotypes and from different cultures. Concomitant medical care may be different. Outcomes may be similar across two studies, but not identical. Endpoint committees code differently. Imaging machines (with availability that is commonly based on hospital contracts and not research needs) have different precision.

The impact of these issues on research results is enormous, has nothing to do with sampling error, and is not erased by changing a p-value threshold from 0.05 to 0.005. Their recommendation focuses on straining out the sampling-error gnat while swallowing the design-inconsistency camel.

However, there is also a philosophical concern. Here is a comment by Jerzy Neyman and Egon Pearson, authored by them in an earnest attempt

---

[*] It is easy enough to change statistical programs to react to thresholds of 0.005 rather than 0.05.

to help others in the early 1930s to understand the heart of significance testing and reproducibility.

> But we may look at the purpose of tests from another viewpoint. Without hoping to know whether each separate hypothesis is true or false, we may search for rules to govern our behavior with regard to them, in following which we insure that, in the long run of experience, we shall not often be wrong. ("On the Problem of the Most Efficient Tests of Statistical Hypotheses." *Philosophical Transactions of the Royal Society* 231, Series A. (1933):289–337)

This is the heart of statistical significance testing. In order to understand the results of such testing, we must (1) give up knowing whether a single experimental result is right for (2) the idea of how correct we are in assessing the long-run experience of multiple research efforts.

But this is not the model of health care research. The Neyman-Pearson approach is fine for flipping coins, but not for clinical trials.

Consider the ALLHAT clinical trial. It recruited over 42,000 patients to study the impact of alternative antihypertensive agents' lipid reductions. This successful but immense and complex study will likely never be repeated on its large scale—similarly for the Women's Health Initiative, which enrolled over 161,000 women.

These studies are one-of-a-kind endeavors. We cannot afford to, as Neyman and Pearson suggest, give up hope in knowing whether each hypothesis is true or false. We must do all that we can to learn if each of these large health care experiments reached the correct result.

Therefore, in order to be sure that we are interpreting the results fairly—that we have the greatest likelihood of an accurate interpretation—we should examine all of the experiment's germane data. They collected a substantial, sometimes overwhelming amount of data to answer the question. That data was designed and embedded to be analyzed and contributory to the research question, not to be discarded simply because it was not included in the alpha calculus of primary endpoints.

Finally, we recognize that different efforts with different designs and executed with different subject populations will produce different results, but perhaps not different conclusions. In order to understand the conclusion in its entirety, one should examine all the data.

Unfortunately, the statistical colloquium (like their colleagues twenty-five years ago) just focused on the p-value.

The metric for reproducibility is not whether clinical trials can produce the same small p-value for the same endpoint. It is instead whether clinical trials designed to answer the same question, recruiting from the sample population base with the same panoply of endpoints each determined with the same precision, can produce results that demonstrate a consistent effect of therapy on the disease. Desirable clinical and epidemiologic reproducibility has little to do with the slavish statistical hypothesis testing results.

This change in the p-value threshold must be resolutely resisted. If it is not, then the p-value, like a tick, will burrow deeper into the tender health care research hide.

But it will not go quietly into the night. We have to force it out.

Conclusions

This is my assessment of where we are with the use of hypothesis testing in clinical research as well as our path to its place.

Neither I nor any of the coworkers that I have been privileged to work with would argue that mathematics has no place in health care research interpretation. When used correctly, it can summarize the findings of complicated research programs.

However, statistical hypothesis testing in general and the p-value in particular fails this test.

By itself (and few people argue that it is useful by itself), statistical hypothesis testing cannot even summarize a simple, single, outcome-measure experiment. It must be accompanied by the effect size, the effect size's standard error, and the confidence interval to provide the assessment of strength of association and also the variability around that strength.

We have much to be thankful for with the introduction of methodologic rigor into health care research efforts that began in the 1950s. We should stay close to these improvements and let their requirement continue to guide our clinical research efforts. Solid, dependable protocols, concordantly executed, are requisites for health care research, but not the p-value.

Experiments now are much more complicated than in the 1950s when the 0.05 rule was first enforced. Clinical trial programs now commonly have multiple treatment arms. They can look at dose response. They can react to a protocol-mandated discontinuation of the treatment arms. They can contain outcomes assessed over multiple time points, multiple outcomes assessed at single, follow-up time point. They contain proper subgroups, complex proteomics, and exploratory analyses.

P-values were simply not designed for this complex environment.

However, unfortunately, rather than set them aside when the research enterprise became complex, the statistical and administrative community doubled down on them. The new research environment excluded subgroup analyses, secondary endpoints, and dose-response relationships (and yes, exploratory analyses) from being quantitatively included in the assessment of the study principally because there was no way statistical hypothesis testing could manage all this.

Rather than discard a constraining metric, they just ignored the complexity of the research program that did not lend itself to the *p*-value, relying on the part of the research program that it deemed interpretable through the type I allocation rule. This is not unlike the hungry man who starves because his weak flashlight does not reveal the feast just beyond of his view.

But let's remember the Japanese proverb. Don't fix the blame. Fix the problem.

We need something better. The following is my idea.

References

[1] Polya, G. *How to Solve It*. 2nd ed. Princeton University Press, 1957.
[2] Kenney-Shafer, L. "When the Alpha Is the Omega: P-Values, 'Substantial Evidence,' and the 0.05 Standard at FDA." *Food Drug Law J*. 72, no. 4 (2017): 595–635.
[3] Moyé, L. A. *Statistical Reasoning in Medicine—The P-value Primer*. 2nd ed. New York: Springer, 2006.
[4] Moyé, L. A. *Multiple Analyses in Clinical Trials: Fundamentals for Investigators*. New York: Springer, 2003.
[5] Bull, J. P. "Historical Development of Clinical Therapeutic Trials." *Journal of Chronic Disease*: 218–248.
[6] Malgaigne, L. F. *Weuvres Completes d'Ambrosise Paré*. Vol. 2. Paris, p. 127. Reported in Mettler, p. 845.
[7] Sutherland, I. "John Graunt: A Tercentenary Tribute." *Journal of the Royal*

Statistical Society 126, A (1963): 537–556.
8. Cochran, W. G. "Early Development of Techniques in Comparative Experimentation." From Owen, D. B., *On the History of Probability and Statistics* (New York and Basal Marcel Dekker, Inc, 1976).
9. Hill, B. "Observation and Experiment." *New England Journal of Medicine* 248 (1953):995–1001.
10. Cochran, "Early Development of Techniques in Comparative Experimentation," ...
11. Israel, Hans, Erich Ruckhaber, and Rudolf Weinmann, Rudolf, eds. *Hundert Autoren gegen Einstein.* Leipzig: Voigtländer, 1931.
12. Fisher, R. A. *Statistical Methods for Research Workers.* Edinburg: Oliver and Boyd, 1925.
13. Fisher, R. A. "The Arrangement of Field Experiments." *Journal of the Ministry of Agriculture* (September 1926): 503–513.
14. Fisher, "The Arrangement of Field Experiments," ...
15. Fisher, "The Arrangement of Field Experiments," ...
16. Fisher, "The Arrangement of Field Experiments," ...
17. Edwards, A. *Likelihood.* Cambridge, UK: Cambridge University Press, 1972.
18. Neyman, J. and E. S. Peason. "On the Problem of the Most Efficient Tests of Statistical Hypotheses." *Philosophical Transactions of the Royal Society (London)* 231, Series A. (1933): 289–337.
19. Pytkowsi, W. "The Dependence of the Income in Small Farms upon Their Area, the Outlay and the Capital Invested in Cows" (Polish, English summaries). *Agri. Res. Inst*, Monograph no. 31 of series Bioblioteka Pulawska (1932). Pulasy, Poland. Wald. A. *Statistical Decision Functions.* New York: Wiley, 1950.
20. Neyman, J. "Outline of a Theory of Statistical Estimation Based on the Classical Theory of Probabilitiy." *Philosophical Transactions of the Royal Society (London)* 236, Series A (1937): 333–380.
21. Neyman, J. "L'estimation statistique traitée comme un problème classique de probabilité." *Actual. Sceint. Instust.* 739 (1938): 25–57.
22. Berkson, J. "Experiences with Tests of Significance: a Reply to R. A. Fisher." *Journal of the American Statistical Association* 37 (1942): 242–246.
23. Berkson, J. "Tests of Significance Considered as Evidence." *Jounal of the American Statistical Association* 37 (1942): 335–345.
24. Fisher, R. A. "Response to Berkson." *Journal of the American Statistical Association* 37 (1942): 103–104.
25. Deming, W. Edwards. *Statistical Adjustment of Data.* 1943.
26. Editorial. "Are Statistics Necessary?" *J. Am. Med. Assn.* 143 (1950): 1260.
27. Ross, Otho B., Jr. "Use of Controls in Medical Research." *J. Am. Med. Assn* 145, 72 (1951): 72.
28. Reid, D. D. "Statistics in Clinical Research." *Annals NY Acad. Sci.* 52 (1950): 931, 933.
29. Bakan, D. "The Test of Significance in Psychological Research." *Psychol Bull.* 66 (1966): 436.

30. Skellam, J. G. "Models, Inference, and Strategy." *Biometrics* 25 (1969): 474.
31. Temple, R. "How FDA Currently Makes Decisions on Clinical Studies." *Clinical Trials* 2 (2005): 276.
32. Carpenter, D. P. *Reputation and Power: Organizational Image and Pharmaceutical Regulation at the FDA.* 2010. 269–97.
33. Goodman, S. N. "Toward Evidence-Based Medical Statistics. 1: The *p*-value Fallacy." *Annals of Internal Medicine* 130 (1999): 995–1004.
34. Marks, H. M. *The Progress of Experiment: Science and Therapeutic Reform in the United States, 1900–1990.* Cambridge, UK: Cambridge Univ Pr, 1997.
35. Porter, T. M. *Trust in Numbers: The Pursuit of Objectivity in Science and Public Life.* Princeton, NJ: Princeton Univ Pr, 1995.
36. Matthews, J. R. "Quantification and the Quest for Medical Certainty." Princeton, NJ: Princeton Univ Pr, 1995.
37. Gigerenzer, G., Z. Swijtink, T. Porter, L. Daston, J. Beatty, and L. Kruger. *The Empire of Chance.* Cambridge, UK: Cambridge Univ Pr, 1989.
38. Hill, "Observation and Experiment," …
39. "Evidence of Cause and Effect Relationship in Major Epidemiologic Study Disputed by Judge." *Epidemiology Monitor* 9 (1988): 1.
40. Walker, A. M. "Significance Tests Represent Consensus and Standard Practice" (Letter). *American Journal of Public Health* 76 (1986): 1033. See also Journal Erratum 76: 1087.
41. Fleiss, J. L. "Significance Tests Have a Role in Epidemiologic Research: Reactions to A. M. Walker" (Different Views). *American Journal of Public Health* 76 (1986): 559–560.
42. Fleiss, J. L. "Confidence Intervals vs. Significance Tests: Quantitative Interpretation" (Letter). *American Journal of Public Health* 76 (1986): 587.
43. Fleiss, J. L. "Dr. Fleiss Response" (Letter). *American Journal of Public Health* 76 (1986):1033–1034.
44. Walker, A. M. "Reporting the Results of Epidemiologic Studies." *American Journal of Public Health* 76 (1986): 556–558.
45. Poole, C. "Beyond the Confidence Interval." *American Journal of Public Health* 77 (1987): 195–199.
46. Rothman, K. J. "Significance Questing." Annals Internal Med. 105 (1986).
47. MRFIT Investigators. "Multiple Risk Factor Intervention Trial." *Journal of the American Medical Association* 248 (1982):1465–77.
48. Feldman, A. M., M. R. Bristow, W. W. Parmley, et al. "Effects of Vesnarinone on Morbidity and Mortality in Patients with Heart Failure." *New England Journal of Medicine* 329 (1993): 149–55.
49. Cohn, J., S. C. Goldstein, S. Feenheed, et al. A Dose Dependent Increase in Mortality Seen with Vesnarinone among Patients with Severe Heart Failure." *New England Journal of Medicine* 339 (1998): 1810–16.
50. MRFIT Investigators, "Multiple Risk Factor Intervention Trial," …
51. Pitt, B., P. A. Poole-Wilson, R. Segal, et al. "Effect of Losartan Compared with Captopril on Mortality in Patients with Symptomatic Heart Failure Randomized Trial—The Losartan Heart Failure Survival Study ELITE II."

Lancet 355 (2000): 1582–87.
52. Multicenter Dilitiazem Post Infarction Trial Research Group. "The Effect of Dilitiazem on Mortality and Reinfarction after Myocardial Infarction." *New England Journal of Medicine* 319 (1989): 385–392.
53. Packer, M. Presentation of the results of the Prospective Randomized Amlodipine Survival Evaluation-2 Trial (PRAISE-2), American College of Cardiology Scientific Sessions, Anaheim, CA, March 15, 2000.
54. UK Prospective Diabetes Study Group. "UK Prospective Diabetes Study (UKPDS) VIII. Study, Design, Progress and Performance." *Diabetologia* 34 (1991): 877–890.
55. Moyé, "*Multiple Analyses in Clinical Trials: Fundamentals for Investigators,*" ...
56. UKPDS Study Group. "Intensive Blood Glucose Control with Sulphonylureas or Insulin Compared with Conventional Treatment and Risk of Complications in Patients with Type 2 Diabetes." *Lancet* 352 (1998): 837–853.
57. Turner, R. C. and R. R. Holman, on behalf of the UK Prospective Diabetes Study Group. "The UK Prospective Diabetes Study. Finnish Medical Society DUOCECIM." *Annals of Medicine* 28 (1998): 439–444.
58. UKPDS Study Group. "Intensive Blood Glucose Control with Sulphonylureas or Insulin Compared with Conventional Treatment and Risk of Complications in Patients with Type 2 Diabetes,"...
59. The Lipid Research Clinic Investigators. "The Lipid Research Clinics Program: The Coronary Primary Prevention Trial; Design and Implementation." *Journal of Chronic Diseases* 32 (1979): 609–631.
60. The Lipid Research Clinic Investigators. "The Lipid Research Clinics Coronary Primary Prevention Trial Results." *Journal of the Amerian Medical Association* 251 (1984): 351–74.
61. Packer, M., M. R. Bristow, J. N. Cohn, et al. "The Effect of Carvedilol on Morbidity and Mortality in Patients with Chronic Heart Failure." *N Eng J Med* 334 (1996): 1349–55.
62. Transcript, May 2, 1996 Cardiovascular and Renal Drugs Advisory Committee.
63. Moyé, L. A. and D. Abernethy. "Carvedilol in Patients with Chronic Heart Failure" (Letter). *N Eng J Med* 335 (1996): 1318–1319.
64. Packer, M., J. N. Cohn, and W. S. Colucci. "Response to Moyé and Abernethy." *N Eng J Med* 335 (1996): 1318–1319.
65. Fisher, L. D, and L. A. Moyé. "Carvedilol and the Food and Drug Administration Approval Process: An Introduction." *Controlled Clin Trials* 20 (1999):1–15.
66. Fisher, L. D. "Carvedilol and the FDA Approval Process: the FDA Paradigm and Reflections upon Hypothesis Testing." *Controlled Clin Trials* 20 (1999): 16–39.
67. Moyé, L. A. "*P* Value Interpretation in Clinical Trials. The Case for Discipline." *Controlled Clin Trials* 20 (1999): 40–49.
68. Antrum, M. Naked Truth and Veiled Allusions. 1901. p. 99.
69. Benjamin, D. J., J. O. Berger, M. Johannesson, B. A. Nosek, E. J.

Wagenmakers, R. Berk, K. A. Bollen, B. Brembs, L. Brown, C. Camerer, D. Cesarini, C. D. Chambers, M. Clyde, T. D. Cook, P. De Boeck, Z. Dienes, A. Dreber, K. Easwaran, C. Efferson, E. Fehr, F. Fidler, A. P. Field, M. Forster, E. I. George, R. Gonzalez, S. Goodman, E. Green, S. P. Green, A. G. Greenwald, J. D. Hadfield, L. V. Hedges, L. Held, T. Hua Ho, H. Hoijtink, D. J. Hruschka, K. Imai, G. Imbens, J. P. A. Ioannidis, M. Jeon, J. H. Jones, M. Kirchler, D. Laibson, J. List, R. Little, A. Lupia, E. Machery, S. E. Maxwell, M. McCarthy, D. A. Moore, S. L. Morgan, M. Munafó, S. Nakagawa, B. Nyhan, T. H. Parker, L. Pericchi, M. Perugini, J. Rouder, J. Rousseau, V. Savalei, F. D. Schönbrodt, T. Sellke, B. Sinclair, D. Tingley, T. Van Zandt, S. Vazire, D. J. Watts, C. Winship, R. L. Wolpert, Y. Xie, C. Young, J. Zinman, and V. E. Johnson. "Redefine Statistical Significance." *Nat Hum Behav* 2, no. 1 (January 2018): 6–10. https://doi.org/10.1038/s41562-017-0189-

70 Johnson, A. L, S. Evans, J. X. Checketts, J. T. Scott, C. Wayant, M. Johnson, B. Norris, and M. Vassar. "Effects of a Proposal to Alter the Statistical Significance Threshold on Previously Published Orthopaedic Trauma Randomized Controlled Trials." *Injury* 50, no. 11 (November 2019): 1934–1937. Epub, August 12, 2019. https://doi,org/10.1016/j.injury.2019.08.012.

What Do We Require of
This New Approach?

A principal problem with the current standard of use of statistical hypothesis testing in health care research is its strict testing regime. Since alpha is commonly prespecified at 0.05 level, there is actually very little error to be distributed among a collection of primary outcomes. If there are too many outcomes the investigators are in the position of either having an enormous sample size, enormous effect sizes, or unreasonably tiny standard errors in order to have a chance for a positive finding through statistical hypothesis testing.

These are considerations that have little to do with biology but are the artifactual consequences of applying statistical hypothesis testing and p-values to health care research, a field of application for which this class of mathematics was not designed.

Instead, these requisites are the price of admission that the clinical investigators must pay in order to be able to simply evaluate whether a treatment in a clinical trial is beneficial or not. Researchers are forced to place the clinical research square peg in the p-value round hole.

A new approach to clinical research assessment should have the following features:

- *It must easily adapt to the multidimensionality of clinical outcomes.* A new approach should take advantage of characteristics of research that may not apply to other scientific fields. For example, there are many ways in which a treatment may be deemed beneficial. It can improve the subject's sense of well-being. It can decrease the likelihood that they will die in a given

39

period of time. It can decrease hospitalization. It can improve measures of morbidity. The subject may exercise longer. They may have improved organ operation (e.g., renal, liver, lung, or cardiac function). Benefit is multidimensional. The same is true for harm. An important feature of a new clinical research evaluation system is that it consider all assessments that the clinical trial makes concerning benefit—not just a subset of them.

- *It must be integrable.* Many estimates of effect in clinical research have their own units. Some are in units of time. Others are percentages; others in milliliters. This new tool must place all these measures of benefit on the same scale so that these benefits can be accumulated. If we are going to assess the impact for each outcome, we must have a way to accumulate these benefits, and a common scale permits that.

- *It must incorporate statistical estimators of different types.* The estimation field has and will continue to make substantial contributions to effect size estimation and its variability. This area will be wholly embraced. Therefore, our new procedures need to include estimators from different types of analyses, be they simple e.g., differences in mean changes, or more complex, such as imputation, mixed model regression, or survival analyses.

- *It must acknowledge estimator variability.* The new tools must not only acknowledge but also take advantage of the observation that there is sampling variability, precision variability, and bias generated by the research design or statistical estimator. The relative risk for the reduction in fatal and nonfatal heart attack may be 0.84, but we must always be clear that while this is the numeral that was computed in the research effort, sample-to-sample variability as well as precision concerns and sometimes bias influence this estimate, and therefore other values are possible.

- *It must be inclusive.* There are many analyses that are conducted in a clinical research effort. There are prospective analyses and retrospective analyses. There are analyses on different subpopulations. Proteomic analyses are of growing importance. We need a tool that will allow us to incorporate these different evaluations. In addition, the new system of analysis must modulate research results by the experience and concerns of

epidemiologists and clinical scientists (e.g., the primacy of prospectively declared outcomes).
- *The tool must be interpretable.* Any accumulation of benefit must be translatable for the research community, physicians, regulators, and patients.
- *Finally, it must be easy to use.* This is the twenty-first century. Investigators should have results produced quickly in easily interpretable tables.

Incorporating all these features will require us to acquire an entirely new perspective for organizing collections of analyses.

It begins with set theory.

The Basics of Set Theory

Motivation for This Work

In health care research, we have collections—collections of questions to be answered, collection of analyses, collections of outcomes, and collections of patients.

Our goal is to develop a construction that permits us to operate on these collections, extracting the cues and messages that they provide about the effect of therapy. Thus, we must have new abilities in manipulating collections of analyses.

Set theory provides exactly the tool kit we need to first allow us to create these collections and then manipulate them. In doing so, we will find that the set theory tools that allow us to combine and disassemble sets will mirror the operations in measure theory that permit us to measure or value these resulting sets.

So we will begin with some basic set theory, followed by an elementary introduction to measure theory.

Those who already understand set theory may skip to the next chapter, where we begin a discussion of measurable functions. However, for the rest of us, before we get enter an exposition into set theory, let's just talk about what you don't need to know to understand it.

- You do not need a degree in mathematics.
- You do not need a statistics degree.
- You do not need a calculus background.
- You do not need trigonometry or geometry.

All you really need is a willingness to understand and an understandable text. If you can bring the former, I commit to provide the latter below.

What Are Sets?

A *set* is simply a collection of objects. These objects can be physical, simply numbers, or metaphysical. A set is defined by its membership criteria.

For example, the collection of US coin denominations is a set. Let's call that set A and define it as {pennies, nickels, dimes, quarters, half dollars, dollar pieces}. If you have twenty-six cents in your pocket as a quarter and a penny, then is {Q P}, where Q denotes a single quarter and P a single penny. Note that the set is denoted by braces ({}).

Each distinct entry in our set is called an *element* in (or of) the set. We denote whether an element s is a member of a set S by the symbol $\in$; $s \in S$ simply means that element s is contained in set S or "s is a member of S."

The order of elements does not matter in sets; sets that have the same elements but are just arranged in different orders are equivalent. Thus, {Q P} is the same as {P Q}. This greatly eases our burden in set manipulation.

Introducing Relationships between Sets

Two sets, denoted by A and B, are equal; $A = B$ if they contain the same elements (again, regardless of order).

Sets are defined by their content; this is equivalent to saying that sets are defined by their membership criteria since it is the membership criteria permitting us to determine if an element is a member of the set or not. It is remarkable that so much mathematical development with so many useful applications can be based on this simple and clear concept—set membership and set comparison.

For example, consider the set of all subjects who are screened for a clinical trial. The membership criteria for this set is that each member has had their demography and comorbidity assessed against the study's

Weighing the Evidence

inclusion and exclusion criteria. We can easily determine if an individual is a member of the set or is not.

To some degree, we are already familiar with sets in clinical trials; we just aren't used to thinking of these research collections that way. For example, the collection of all subjects accepted into a particular clinical trial is a set (the membership criteria is simply acceptance into that clinical trial). Similarly, individuals who are between forty to seventy years of age in that clinical trial also comprise a set; in fact, it's a subset of those individuals who are in the trial.*

As we just saw, sets can contain other sets. These contained sets are known as subsets. Thus, if S is the set of all subjects in a clinical trial and F is the set of females in the trial, then the statement that "the collection of females in the clinical F is part of S ($F \in S$) is true. Females represent a subset of S. We can also say that "F is contained in S" or $F \subset S$. Another true statement is that "S contains F" or $S \supset F$

It will be quite useful for us to declare that a set has no elements. A set that has no elements is the "null set," denoted by { }, or more commonly $\varnothing$. Thus, if a clinical trial carried out no imputation analyses, its set of imputation analyses I is said to be null; we may write I = { } or $I = \varnothing$, denoting that the set of imputation analyses is empty or null.

Set Operations

One of the reasons that numbers function so effectively and practically in our society is because we can easily manipulate them. We can add to them, subtract from them, and compare them. We will need that same facility when working with sets. The principal set operations we will be working with are unions, intersections, and complements.

Unions

Let's begin with the set P that contains all subjects included in a health care research study. If the study has n subjects, then the set P contains the same n elements. We can already think of subsets of P such as the subset of individuals with LDL cholesterols greater than 175 or

---

* In clinical trial methodology, we commonly think of this collection of subjects as a subgroup, but this collection also meets the definition of a set as well.

the subset of individuals who were exposed to potassium sparing diuretics.*

From the set of all subjects in the study P, let's define A_{40} to be the subset of all subjects greater than 40 years old and M as the set of all males. We can describe the subjects who are in either set as those subjects who are either greater than 40 or are male. This, we say, is the union of A_{40} and M or $A_{40} \cup M$.

This union combines the elements from both sets into a new set. So since we know that this union contains all subjects who are greater than 40 years old regardless of their gender and addition contains all males regardless of their age, then we know that $A_{40} \subset A_{40} \cup M$. We also know that $M \subset A_{40} \cup M$.

Now we expect that there may be redundancy in this union. Males greater than 40 years old are in both A_{40} and M However, the union counts them once and only once.

Working with unions requires practice. A good rule of thumb is to be absolutely clear about the set membership.

For example, does $A_{40} \cup M = P$, or is it merely contained in P? It depends on the recruitment for the study. If the study only recruited males, then the union is equal to P. If the study only recruits subjects who are greater than 40 years of age, then $A_{40} \cup M = A_{40} = M = P$.

However, if the study contains one female less than or equal to 40 years of age, then $A_{40} \cup M \subset P$. If we don't know the female's age, then we can say $A_{40} \cup M \subseteq P$., which means that the set $A_{40} \cup M$ is contained in or equals the set of all subjects in the study P.

Intersections

Now continuing with the same example of the sets A_{40}, M and P, we now ask, who are the individuals in

---

* Of course, it's possible that each of these subsets might be null if it has no subjects with these characteristics.

both A_{40} and M? If we let ω represent an individual P, (i.e., $\omega \in P$,), then what are the characteristics of this individual ω when $\omega \subset A_{40}$ and $\omega \subset M$?

These are individuals in the study who are male and over 40 years old. We call this set the intersection of A_{40} and M symbolized as $A_{40} \cap M$.

This understanding of intersection is all that we need to say that $A_{40} \cap M \subseteq A_{40} \cup M$. The statement is true because if an individual is in both sets, then that individual is in either of them; and therefore, they are in the union. However, an individual could be in the union (e.g., a male who is 29 years of age) and would not be in both of them.

One way to remember the construction of intersections is that members of intersections must be in each and every set that makes up the intersection, i.e., if $\omega \in A \cap B$, then $\omega \subset A$ and $\omega \subset B$. However, a member of a union only needs to be a member of at least one of the sets of the union.

Complements

Finally, we have the notion of a complement. The complement of a set B is the set of all members who are not in the original set. The complement of B is denoted as B^c. In our current example, A_{40}^c is the set of individuals who are less than or equal to 40 years of age, and of course, M^c is the set of all females in the study. Note that $A_{40} \cap A_{40}^c = \emptyset$ since individuals must be in one set or the other, but not both; this property is termed mutual exclusivity. Also, $A_{40} \cup A_{40}^c = P$ since an individual must have an age and that age is either less than or equal to 40 or greater than 40.

Can we subtract sets?

What would $A—B$ look like?

Let's think about what $A_{40} \cap M^c$ must be. We know that $A_{40} \cap M$ is the set of all males greater than

40 years old. Taking the intersection, not with M but with M^c, leaves us the set of females greater than 40 years old. Another way to think of this is that the set $A_{40} \cap M^c$ is the set of all individuals greater than 40 years old with the males "removed." This combination of operators acts like we would expect the operator $A_{40} - M$ to operate. So although one cannot technically subtract sets, the combination of the intersection and complement operation allows us to accomplish exactly that.

Those who want to examine some more complicated manipulations in set theory can proceed to the intermediate set theory discussion in the following chapter. For the rest of us, let's consider what we can do with the operations that we have considered thus far.

Venn Diagrams

It does not take much imagination to understand that set operations and manipulations can become quite complex, e.g., $(A \cup B^c) \cap (D^c \cup C)^c$ for arbitrary sets A, B, C, and D. Venn diagrams are particularly useful in being able to see and study the impact of these operations (figures 1 and 2).

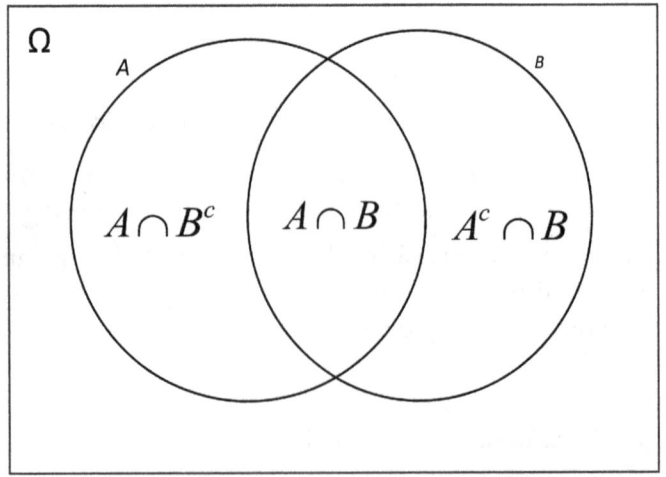

Figure 1. Demonstration of unions, intersections and complements using 2 nondisjoint sets A and B.

Weighing the Evidence

Let's consider some arbitrary sets A and B. From figure 1, we can see that $A \cup B$ can be constructed from overlapping sets, e.g., simply $A \cup B$ or from three nonoverlapping, disjoint sets, $A \cap B^c$, $A \cap B$, and $A^c \cap B$.*

We can also see that $A = (A \cap B) \cup (A \cap B^c)$., observing that $A \cap B$ and $A \cap B^c$. are disjoint sets.

The union operation can produce some interesting results (figure 2).

$A \cup B = A;\ A \cap B = B$

$A \cup B = \{A, B\};\ A \cap B = \emptyset$

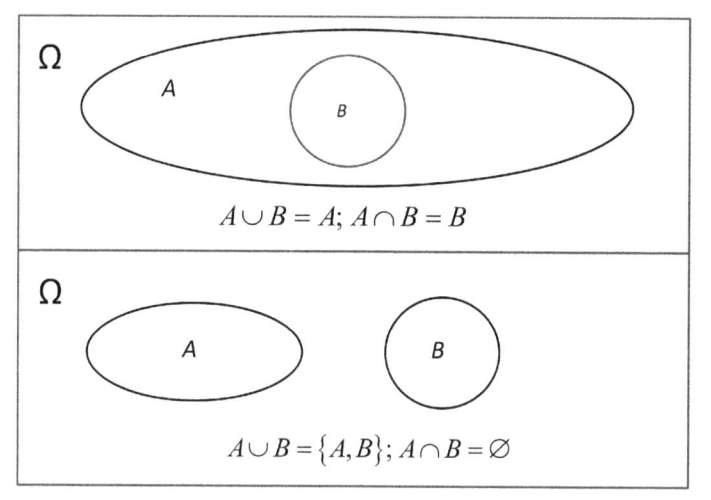

Figure 2. The union set $A \cup B$ depends on the degree of overlap of sets A and B, depicted by $A \cap B$.

For example, when B is wholly contained in A, i.e.,
$B \subset A$, then $A \cup B = A$, and $A \cap B = B$. In the case where they are disjoint, then we can see that $A \cup B = \{A, B\} : A \cap B = \emptyset$.

The distribution law shows us how to work with three sets.

---

* Note from our earlier discussion that we can consider the set $A \cap B^c = A - B$, and $A^c \cap B = B - A$.

49

Distribution Law of Sets

$$(A \cup B) \cap C = (A \cap C) \cup (B \cap C)$$
$$(A \cap B) \cup C = (A \cup C) \cap (B \cap C)$$

The use of parentheses permits us to avoid any ambiguity in the order of operations. For example, the operation $(A \cup B) \cap C$ contains every member that is in both A and C, or every individual who is in both B and C. while the ambiguous statement $A \cup B \cap C$ could be the former or could be the union of every element in set A with elements that are in both sets B and C We always conduct the operations in parentheses first. If parentheses are nested, we begin from the inside and work our way out.

Another useful tool is De Morgan's law.

De Morgan's Law

$$(A \cup B)^c = A^c \cap B^c : (A \cap B)^c = A^c \cup B^c.$$

De Morgan's law gives us a way to manage set complements. Taking a minute to think them through provides some insight. For example, the second law begins with $(A \cap B)^c$. This is a collection of individuals who cannot be in the intersection of sets A and B Thus, they can be in set A alone, B alone, or neither of them. This is precisely those who are not in A nor those who are not in B as De Morgan's law states.

Unions, intersections, and complements become more complicated when we consider the circumstance of three sets (figure 3).

Taking the unions, intersections, and compliments of sets is a straightforward way for sets to generate other sets, which themselves, through the same operation, produce additional sets. Essentially, these three operations put us in the set-generation business.

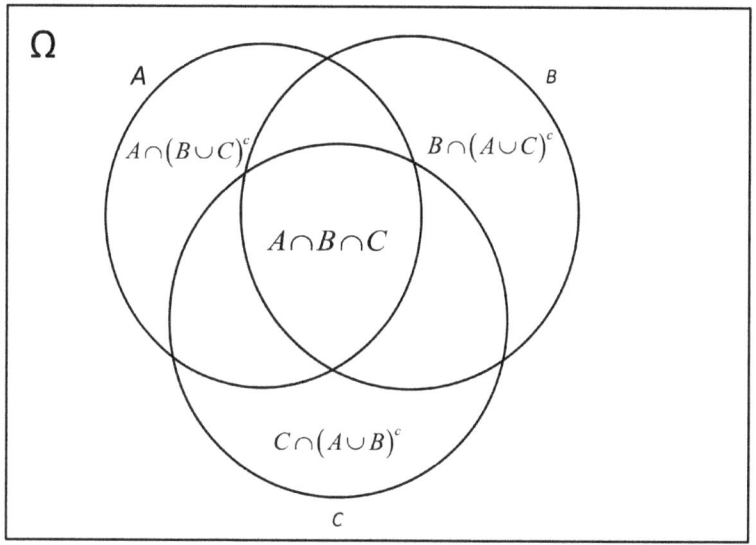

Figure 3. Demonstration of unions, intersections and complements using 3 nondisjoint sets *A* and *B*.

Set Generation and σ-Algebras

As we have seen, the set operators union, intersection, and complement generate new sets that are related to but are different from the original sets. The numbers of these new sets can be far larger than we might expect. For example, for $\Omega = \{A, B\}$, we can generate sets as follows:

$$A, B, A^c, B^c, A \cap B, A^c \cap B, A \cap B^c, A \cup B, A^c \cup B, A \cup B^c$$
$$(A \cup B)^c, (A \cap B)^c, \emptyset$$

We generated as many as 13 from the original two sets. If $\Omega = \{A, B, C\}$, then we could generate many, many more sets. In fact, if there are *n* elements in Ω then the number of subsets generated by these three set operations is greater than *n*!

This set-generation feature is central to our future use of measure theory. Specifically, all the sets that are generated by these operations of union, intersection, and complement we will call the *sigma algebra* or σ-algebra. A σ-algebra is nothing more than a collection of subsets of the set Ω (we will designate that collection of subsets as Σ), which

51

follows certain rules of inclusion precisely satisfied by taking unions, intersections, and compliments.

Creating the σ-algebra is straightforward; we start with a collection of sets, then generate from that collection the null set and every possible combinations of unions and complements.[*]

The precise definition of the σ-algebra, Σ, of subsets Ω is the following collection of subsets:

a) The null set is a member of Σ, $\emptyset \in \Sigma$.
b) If the set $A \in \Sigma$ then $A^c \in \Sigma$.[†]
c) If a countable[‡] number of sets $A_1, A_2, A_3, \ldots A_n$ are contained in Σ, then $\bigcup_{i=1}^{\infty} A_i \in \Sigma$.

This formal definition implies that intersections of sets are members of Σ as well. So defining a σ-algebra in terms of unions and complements also implies that this σ-algebra must contain their intersections as well.[§] In the end, we have to simply keep in mind that a σ-algebra is nothing more than all the subsets of elements contained in Ω.

Examples of σ-algebras. Consider a collection of five DVDs with unique titles T_1, T_2, T_3, T_4, T_5. The original set of them is simply $\{T_1, T_2, T_3, T_4, T_5\}$. We can construct the σ-algebra Σ as

$$\emptyset, \{T_1, T_2, T_3, T_4, T_5\}, \{T_1\}, \{T_2\}, \{T_3\}, \{T_4\}, \{T_5\}, \{T_1^c\}, \{T_2^c\}, \{T_3^c\}, \{T_4^c\}, \{T_5^c\},$$
$$\{T_1 \cup T_2\}, \{T_1 \cup T_3\}, \{T_1 \cup T_4\}, \{T_1 \cup T_5\}, \{T_2 \cup T_3\}, \{T_2 \cup T_4\}, \{T_2 \cup T_5\},$$
$$\{T_1 \cup T_2 \cup T_3\}, \ldots$$

---

[*] As an example, consider a set of tracks $T = \{t_i\}, i = 1\ldots n$ be all of your music tracks. Then the σ–algebra is all of the possible playlists you can construct from these tracks (including the playlist that contains no tracks at all!).

[†]

[‡] By *countable*, we mean there is an infinite number of sets that correspond with those numbers. One can be begin with 1 and proceed to infinity without missing any of the sets.

[§] Assume A and B are contained in Σ. Then A^c and B^c must be contained in Σ. But their union $A^c \cup B^c$ must also be in Σ as must their complement $(A^c \cup B^c)^c$, which, by De Morgan's law, is $A \cap B$.

and on and on, continuing to build this collection of sets up through the unions, intersections, and complements. From a set with five elements, the σ-algebra becomes very large. By containing all unions, intersections, and complements of set elements, the resulting collection of subsets can be very rich. It all depends on the elements in the original set.

Painting. A particularly useful way to consider the role σ-algebras would be in painting. Suppose one has a gallon of red paint. Then the combinations of colors generated from it is very small—essentially no color (corresponding to the null sest) or the color red. Thus, the σ-algebra consists of only two elements.

However, suppose you now add three additional gallons of different colors—one each for black, blue, and yellow. The σ-algebra of all four colors is still all the combinations of colors that can be generated by combining them, but because the original set is larger, the collection of subsets is very rich.

The oranges, crimsons, purples, grays, teals, pinks, apricots, lavenders, boysenberries, and so many others are all members of a huge mixture of new colors produced by combinations of the original set. Since the original set was richer, the σ-algebra has exploded.

Subjects. As a final example, let Ω equal the number of subjects in a clinical research program. Here, the σ-algebra is all the subgroups of this population. This is each subject individually, then all subjects taken two at a time, then taken three at a time, and so on. The total number of ways to gather these individuals into different distinct collections is immense.

Of course, the number of subgroups actually analyzed is infinitesimal compared to the total number of subgroups actually used in a clinical trial since only a small number are phenotypically meaningful.

Why We Need σ-Algebras

The concept of a σ-algebra is meaningful for us because it will be the basis of an entirely different class of functions.

We typically think of functions as operations that map one number to another number, such as $y = x^2$. However, our new interest will require that we not just map numbers to numbers, but sets to numbers.

This will be a new matter for most of us. We will map items to numbers and then collections of items to numbers. And the greater and more diverse the items in the original Ω the richer the σ-algebra of events, which will be the argument of our function.

For example, envision a system in which all analyses conducted in a clinical research enterprise are collected. We define this set of analyses Ω and then identify all possible sets and subsets of Ω This resulting σ-algebra Σ of sets can be mapped to specific numbers based on their common traits or characteristics. For example, analyses which produce estimators that suggest that the exposure is beneficial can be mapped to one number, while those that suggest harm can be mapped to another.

However, in order to keep good structure and order, we need to follow some mathematical rules in these set mappings. These rules help us to define measurable functions and then measure.

Elementary, Set, and Measureable Functions

Here we will describe the development of set functions that can be of use to us in health care research. We will do so with some very simple examples. This will set the stage for the definition of a measurable function and then of the concept of measure.

Let's continue where we left off. Assume that we have a sample space or σ-algebra complex denoted as (Ω, Σ) and that we have members ω_i, such that $\omega_i \subset \Sigma$. Recall that the σ-algebra Σ can be explosive in size. However, our goal is not to simply tabulate elements of sets. Ultimately, we want to assign values to these elements and then values to sets. In order to do that, we need a special type of function called a set function, and its simpler version called an elementary function.

Let's start with the easiest—an indicator function. It is denoted by $1_{[]}$. This function is defined as either 0 or 1 depending on the condition that is defined in the bracket of the subscript. For example, suppose that Ω is the set of all statistical analyses conducted in a clinical research endeavor and Σ is its σ-algebra. Define $f(\omega_i) = 1_{[\omega_i \text{ is a } t-test]}$. Then in order to assign the value of $f(\omega_i)$, we simply inspect ω_i to determine if it is a t-test. If it is, then $f(\omega_i) = 1$. It not, then $f(\omega_i) = 0$.

Note that we are using this indicator function to convert the presence of a condition for ω_i (which, in this case, is not a number) to a number.*

---

* Also, in order for us to even assign the number, the property (in this circumstance, that property is whether the analysis is a t-test or not) had to be available for inspection. This availability is an essential feature of a measureable function, which we will discuss later.

Also, observe that this function assigned a value to an element of the set (as opposed to an entire set of analyses). We call such a set function an elementary function and denote it by $e(\omega_i)$ In our example, we would write $e(\omega_i) = 1_{[\omega_i \text{ is a } t-test]}$. An elementary function is a special set function that assigns a number to single element of a set ω_i that is contained in our Ω, Σ.

For example, if (Ω, Σ) is the sample size and σ-algebra of all subjects randomized to a clinical trial, then we might create the elementary function $h(\omega_i)$ to determine if the ith subject is Hispanic. It would operate on each and the element in Ω Specifically, we would write this function as $h(\omega_i) = 1_{[\omega_i \text{ is Hispanic}]}$. This function essentially inspects each subject's ethnicity and assigns that subject the value 1 if they are Hispanic and 0 otherwise. We can imagine other such functions for age, gender, and combinations of demographic factors.

The elementary function is a special indicator function that maps a single element of a set to either 0 or 1; its domain is a singleton element ω_i of a set.

However, it must also have the characteristic that the property inspected by the elementary function must be a property that is available to be inspected.

Measurability

Measurability is a critical concept in the development of set functions. However, it is a concept that can be easily explained for nonmathematicians.

Measurability is a property of a function. A function is either measurable or nonmeasurable. Our work will concentrate on measurable functions.

Measurable functions have two properties.

The first is that the function itself must return either zero or a positive value. It cannot be negative.* This property is automatically handled by the 0-1 definition of the indicator function. This is straightforward.

---

* We will compensate for this limitation in value by showing that we can premultiply by a sign, such as $-h(\omega_i)$

The second property involves the inspection of ω_i. For example, the function $h(\omega_i) = 1_{[\omega_i \text{ is Hispanic}]}$ inspects for the ethnicity property. If members of the set Ω have that property to be inspected, then the set function is a measurable function on Ω

A function that would not be measurable on Ω would be $a(\omega_i) = 1_{[w_i \text{ is an Android phone user}]}$. The type of smartphone that a subject in a clinical trial possesses (or whether they actually have one) is not available in clinical trial databases. Thus, the function cannot operate because the property that it wishes to inspect is not available.

As seen in this example, there is no use for us to develop nonmeasurable functions for our application to clinical trials. Thus, the functions that we develop will be measurable on (Ω, Σ). However, the availability of this property we must always keep in mind.

For example, if (Ω, Σ) is the set of all analyses conducted in a clinical trial and $r(\omega_i) = 1_{[w_i \text{ is a regression analysis}]}$, then by definition, $r(\omega_i)$ is nonnegative and measurable.

As another example, we can define an elementary function for an analysis as 1 if the analysis contains patients with diabetes, and 0 if it does not. This would be symbolized as $e(\omega_i) = 1_{\omega_i \text{ contains diabetics}}$. If the members of ($\Omega, \Sigma$) contain information about the morbidity of patients in the clinical trial (a likely set of circumstances), then $e(\omega_i)$ is measureable.

If we have a collection of participants in a clinical study and we want to know how many of them returned for a visit six months into a study, we could define the set Ω as the set of all individuals ω_i in the study and the set B as the event that a participant returned for the six-month visit. Then if we define $e(\omega_i) = 1_{[\omega_i \subset B]}$ for each subject ω_i in the study, then each ω_i is mapped to either 0 or 1 depending on whether the individual returned for their six-month visit. This is measurable or not based on whether each element ω_i contains the follow-up information. For example, if (Ω, Σ) is created at baseline, the function is not measureable because follow-up information is not yet available. However, if (Ω, Σ) is created at the end of the study when follow-up information is available, then $e(\omega_i)$ is measurable.

Lem Moyé

Broadening the Elementary Function

At this juncture, we are comfortable with the notion of an indicator function and at least know how to determine if it is measurable or not. Let's now introduce the concept of the more general set function.

A set function maps a set (not just an element of a set like an elementary function does) to 0 or 1. Unlike the elementary function, the set function's argument is a more general set (that can contain more than just one element). Thus, set-indicator functions are related to but are different from elementary-indicator functions. An elementary-indicator function maps an individual element of a set to 0 or 1 based on whether that element is in another set. The broader set function assigns 0 or 1 to the set itself.

As an example, let's return to our clinical trial example where (Ω, Σ) contains the individuals in a clinical trial at baseline. Let the set of all males be M. Since M is a subset of Ω, then $M \subset \Sigma$. Now consider two functions. The first will be the elementary function $e(\omega_i) = 1_{[\omega_i \subset M]}$ assigning 1 if the ω_i^{th} individual is male and 0 if not. The function $e(\omega_i)$ is our standard elementary function.

Now let's create a new function g. Its argument is any subset A of individuals in Ω and $g(A)$ is defined as the number of males in set A.

Both $e(\omega_i)$ and $g(M)$ are related to each other since they each involve the set M; however, they use M differently. The function $e(\omega_i)$ uses M to identify a characteristic of the individual ω_i and assign a value to ω_i based on the presence or absence of that trait (in this case, the characteristic of male gender). It operates on any $\omega_i \in \Sigma$.

However, the function g does not concern itself with a trait of a single member of a set but of the entire set itself.* So while $e(\omega_1, \omega_2, \omega_3)$ cannot be computed, $g(\omega_1, \omega_2, \omega_3)$ makes perfect sense. These two functions are related by $g(M) = \sum_{i=1}^{n} e(\omega_i) = \sum_{i=1}^{n} 1_{[\omega_i \subset M]}.$ †

---

*   Of course, one can make the argument that the individual ω_i's are sets themselves, but the distinction here is whether the argument of the function can have more than one element.

†   We will see that in general, this can be written as $g(M) = \int e(\omega_i)$.

We will find in public health problems that it is usually easier to start with the elementary function when building a complicated simple function and then to convert it to a more general set function.

We can also easily see that functions constructed from measurable elementary functions are also measurable. As an example of why this is true, suppose that we define Ω as a collection of individuals on whom demographics including age and race are available, and Σ is the σ-algebra of these individuals. Now consider the function $k(\omega_i) = 1_{age \leq 45} + 1_{Hispanic}$. For any value of $k(\omega_i)$ one can find the age and ethnicity of individuals so that the function value can be assigned. The key to this is noting that since the elementary functions are measureable, their sums and difference are measurable as well.

Summary

Let's take a breather for a moment and figure out where we are.

We began with a very basic definition of collections or sets of items and simple ways to combine and compare two of them. This led to the construction of a collection of sets reflecting all of the many possible ways the original sets elements could be combined.

One example that we have used in clinical research and will continue with is the collection of all analyses conducted. We call that collection of all analyses the set Ω.

From Ω, we have generated the large collection of sets reflecting all possible combinations of analyses. This huge collection of sets are gathered together into a σ-algebra we term Σ

Commonly Ω and Σ are combined into their own collection (Ω, Σ) With this as a foundation, we defined the notion of a set function, which simply maps members of Σ to a real number. A measurable function is a set function whose criterion for inspection is a property of each element in Σ (i.e., mapping criteria is available in Σ) and whose possible values are only positive.

There is a huge number of measurable functions on the (Ω, Σ) analysis space. For example, if $\{\Omega, \Sigma\}$ represents the set of all hepatocytes in a subject's liver, then one possible measure $v(A)$ might return the ALT enzyme content of that set, and a wholly separate measure $\varsigma(A)$ could return the AST consent of that same set A.

In addition, the measure could simply be a 0-1 dichotomous measure, e.g., does the set A contain any bilirubin? For example, the measurable function $g(\omega_i) = 31_{\omega_i \subset S}$, where S is the set of all subgroup analyses, is a measurable function on our (Ω, Σ) foundation.

However, we will need to discipline ourselves to identify the set of measurable functions that will be the most helpful and that also reflects the research circumstances. This will provide a foundation that is custom-tailored for health-care research analyses.

Now we will take the measurable approach one (final) step further and identify a way to assess the value or the contribution of a set (e.g., a collection of research analyses). This is determining the measure of a set.

Measure and Its Properties

Measure theory is based on the idea that the content of a set can be mathematically assessed and valued. At this point in our development, we can manipulate sets; now we begin to study how we can measure them.

It is tempting to think that we can already do this with measurable functions since, after all, these map the content of a set to a nonnegative number. However, we will see that measure requires more than just assigning a number. There must also be an induced relationship between the value of different sets.

For example, if a first set is wholly contained in a second set, shouldn't the value of the second set be at least as great as the value of the first set? Measurable functions do not guarantee this.

On the other hand, we will also see that we are already in the habit of providing content for sets although we are not used to thinking of it that way.*

Here we will spend some time providing the background for measure that we will be able to use in health care research, our ultimate goal.

The set theory, however, does get somewhat thick. To help you with its absorption, I have created two paths.

The elementary path is for those who have absolutely no background in measure theory. This path will not make you an expert, but you will learn what you need to know in order to understand measure theory's basic properties and to apply it in health care research. To travel this path, continue reading this chapter.

---

* One such example is assigning probability to events. Probability is just one of many different ways of assigning measure to sets or events.

The intermediate path delves into the mathematical underpinnings of measure theory, skipping the introductory, nonmathematical preamble. There are substantial set theory findings on this tack. If this is for you, go straight to the "Measure and its Properties" chapter.

The advanced path dives straight into the development of a new measure for clinical research. You may turn to "An Interlude", proceeding from there.

If you are unsure, or need a refresher in thinking about measure theory, then simply continue reading.

Elementary Path

So let's begin with what measure theory can do, by beginning with a kidnapping.

> "Don't make a sound, put your hands in your pockets, and get in the back of that car."
>
> Your neck is yanked by a hand whose vice grip forces your head into a paper bag smelling sweet-sick like old milk. Shoved backward into a car seat, your arm scraping the rough edge of the cold, torn leather, you hear the door close and feel the car peel away from the curb.
>
> After several abrupt turns, other car horns blaring in protest at this vehicle's jerking movements, you lose track of the sequence of lefts and rights.
>
> No idea where you are.
>
> The car screeches to a halt, the door opens with a bone-chilling squeal, and you are pulled out and shoved into a hot and humid room. After getting pushed down into a chair, the smelly paper bag is removed from your face.
>
> Blinking your eyes back to focus, you see before you money.
>
> Piles of money.
>
> Heaped to overflowing on a round table. Shaking your head to get away from the dirty bag's smell, you

inch closer to see that although it is money, it doesn't look exactly like money.

"Yeah, that's right," scratches the reedy voice of a woman from behind. "All kinds. Dollars, pounds, shekels, drachma, francs, rubles, deutsche marks. We even have a few thousand native beads in there. We need to know how much it all is worth. Why do you think we brought you here?"

For us, counting all that money in its various forms goes to the heart of our use of measure theory.

What Is Measure Theory

Measure theory, at its heart, is the science of measuring the accumulation of things. Sometimes the rules of accumulation are complicated. Other times (e.g., simple counting), the gathering is easy to follow. In any event, the ideas are always straightforward.

Measure theory and its generalities envelops much mathematical content; probability is one of its subfields. Developed by the French mathematician Lebesgue, and the Russian probabilist Kolmogorov in the 1920s, measure theory is based on set functions, those functions that we discussed that map sets to numbers.

Ultimately, we take a collection of sets, assess each set's value, and accumulate that value over each of the sets. In the end, we have the accumulated total value of all the sets. And since sets can be arbitrary, we have substantial freedom in choosing the sets that are to be valued. However, we have to follow certain mathematical rules in this valuation.

Accumulation

For our purposes, measure theory focuses on the process of accumulation (gathering together or rounding up or congregating) the value of items.

We already know how to combine items if we think of set elements as the items. The operations of unions, intersections, and complements are the manipulations by which set elements are combined or gathered into new sets.*

This accumulation process is the heart of measure theory. What can appear complicated about measure theory is that the accumulation process may be complex, using different procedures to measure different quantities. However, what appears at first glance to be a complicated myriad of arbitrary rules turns out upon further inspection to be precisely the combination of procedures required to accurately accumulate the required quantity. Let's start with some easy examples.

> *Example—Measuring Wealth*
>
> Consider the task of measuring the wealth accumulation for a typical five-year-old US boy over the course of his life. At the beginning of the process, everything that this five-year-old possesses (e.g., clothes and toys) is purchased by his parents. The wealth that he has truly earned comes solely from his own weekly allowance or small financial gifts, an allowance that can be measured by simply counting up the value of the coins that he is either paid or that he occasionally finds (e.g., from the tooth fairy). Since collecting coins is the only way by which he accumulates his own independently earned wealth, we are content to only count the value of coins. This is easy—we know how to *measure* coins.
>
> However, as the boy grows, he starts to accumulate wealth in other ways. One natural development is to earn not just coinage but also paper money. This change poses a dilemma for our measurement of his accumulated wealth. If we continued to base our measure of his wealth solely on the value of coins, we would miss an important, new (and greater) source of

---

* The operation of computing the probability of an event by (1) showing how that event is the "combination" of simpler events whose probability is easy to find and (2) using the rules of set operations to reconstruct the event of interest from the simpler sets and (3) then applying the rules of probabilty, operation by operation, to build up the probability computation of the more complicated event from the simpler set comes from measure theory.

Weighing the Evidence

his earned wealth, and consequently, the estimate of this earned wealth would be in error.

We therefore very naturally alter our wealth-counting mechanism to now include a new counting procedure—the accumulation of the value of paper money.

Note here that the tools used to count money have changed (from coin value to a combination of coin value and paper-money value), but not the goal of measuring his accumulated wealth. Our counting mechanism had to flexibly adapt to the new economic situation if it was to remain accurate. Since accuracy is the key, we change the manner in which we count, but we remain true to the process of assessing wealth accumulation. We could say that we adjusted our "measuring tool" to now count not just coins but dollars as well.

Additional changes in how we accumulate wealth are required as our subject prepares to go to college. How should the process adapt to the mechanism of the boy (now a young man) who uses his own money to buy a car?

Simply continuing to merely count coin and paper-money value as a measure of his independently acquired wealth clearly introduces inaccuracies. As he acquires smart devices, computers, cars; obtains a paying job; invests in stocks and bonds; buys and sells homes, etc., our accumulation process, which started with simply recognizing coin denominations, must adapt repeatedly to include these new forms of wealth. Again our rules of accumulation had to adapt to the changing, increasingly complex reality of the circumstances.* Yet the goal of the process remains the same––the *measure* of the man's wealth.

This is a complex process that produces, in the end, a fairly complicated function. However, while that function may not be recognizable at first glance,

---

* The consideration of depreciation of these material assets over time is yet one more easily added addition to our increasingly complex rules in estimating this individual's wealth.

we understand how it was developed and can use it to measure the individual's worth.

This we have done by simply creating sets where the assessment of wealth is the same type of value for each element (coin set, property set, stock set, etc.), and then we assign the right value to each set and subsequently accumulate the value. Thus, the sets have to be evaluable by the measure, and we may need to apply a different measure (or value assessment) to each set

Example—Music Tracks

Many people now manage their songs (or tracks) digitally. Suppose an individual with several thousands tracks wishes to get a sense of this collection's value or worth. How could they do this?

One way would be by simply counting the number of tracks, beginning with perhaps the oldest and moving to the latest downloads and increasing the count by one for each distinct track. This is simply and naively "counting measure." In the end, one knows the total number of tracks, and in a sense, one has *measured* them.

If we call this collection of tracts A, then the value of the tracks might be as easy as $V(A) = \sum_{i=1}^{n} \mu(i) = \sum_{i=1}^{n} 1 = n,$ where the measure of the ith track, $\mu(i)$, is simply 1, and n is merely the number of tracts in the collection of music A

However, another equally valid way to proceed is to place a value on each track—for example, the number of times that track has been played. Many tracks may have never been played, while others may have been played hundreds of times.

In this example, one accumulates the "size" or "measure" by adding not the track but the number of plays that it has been played. This will lead to a different measure of the music collection. Let $\eta(i)$ be the number of times that the ith track has been played.

Then there's this type of accumulation, $N(A) = \sum_{i=1}^{n} \eta(i)$, which is the accumulation of plays and $N(A)$ is the measure or the value of the music collection A.

A third measure would be the duration of the track in time. Here, one simply accumulates or sums the length of each track in the end coming to a time (e.g., 17.7 months). $D(A) = \sum_{i=1}^{n} d(i)$, would be the accumulation duration playing time for the music collection A.

Which of these measures is right?

They are all right. Each (total tracks, number of plays, and total time) is legitimate because each value or measure is based on or can be traced back to a measurable characteristic of the music collection. However, each measure is different because it emphasizes a different property of the music.

Example—Clinical Research Reimbursement

As a final example, suppose you are in charge of making payments in a clinical study that will follow subjects over a period of time.[*] The clinical centers that recruit these subjects will, of course, incur substantial cost as they see and examine each patient, draw blood work, and obtain modern (and expensive) imaging over the course of the study.

Assume that each study patient will be seen six times over the course of the research. How should the coordinating center reimburse the centers for their costs?

One idea (plan A) reimburses the centers directly in accordance with the way that costs were incurred—in this case, making equal payments of 16.7% of the total cost on each of the entire six months so that by the conclusion of the study, the clinics have received 100% of the payments.

---

[*] This is based on an example provided by Rachel W. Vojvodic, MPH.

However, plan B assigns dollars differently. It provides 60% of the cost divided equally over the first two visits, then 10% during the remaining four visits. This front-loading of cost permits the clinical center to expand their research team early in the study to provide more accurate and timely patient throughput and data transmission.

Alternatively, plan C back-loads costs, paying 10% of the total cost for each of the first five visits, then 50% for the last visit. This adds an important financial incentive to the scientific motivation of clinical centers to follow study subjects to the end of the research.

Each of these plans provides total cost disbursement at the conclusion of the study; however, the distribution of costs is different (figure 1).

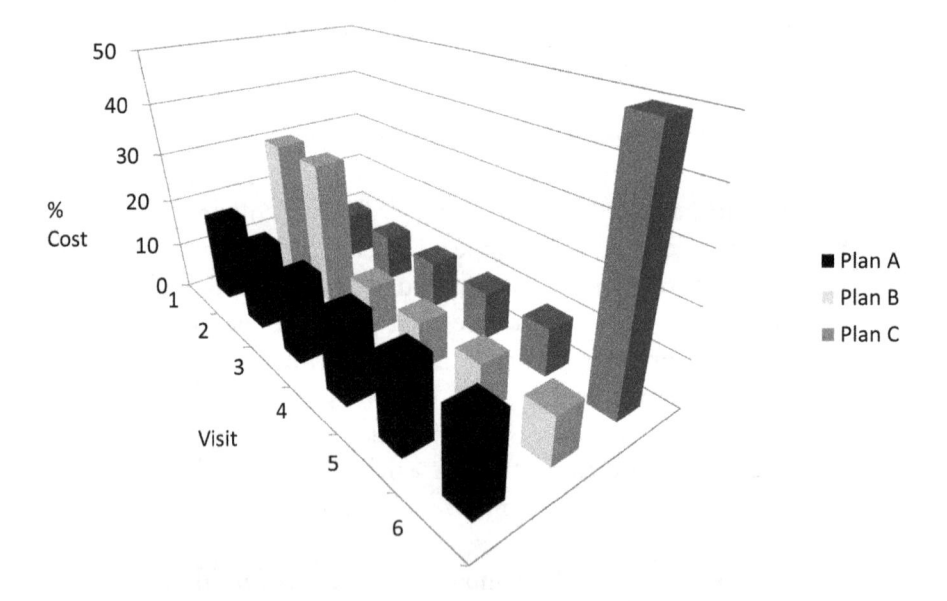

Figure 1. Three different plans for cost reimbursements over time in a clinical study.

Suppose that we want to compute the cost reimbursement for the first three visits of each plan. Plan A reimburses approximately 50% of the total patient care cost during this period. Plan B reimburses 60% during that period of time, while Plan C reimburses 30%. Now define the cost for a visit as the measure of that visit. The costs or measure of each of these plans during the first three visits is different. The total measure over the six visits is the same or 100%.

If we characterize the visits as $V_1, V_2, V_3, V_4, V_5, V_6$, then we can go even further and define measure μ as the reimbursed cost for each visit.* So the cost for V_1 as $\mu(V_1)$ and the cost or measure of visit 1 under plan A is $\mu_A(V_1) = 16.7$. Then the system of cost or measure of both V_1 and V_2 is

$$\mu_A(V_1 \text{ and } V_2) = \mu_A(V_1) + \mu_A(V_2) = 16.7 + 16.7 = 33.4.$$

We can also see that $\mu_B(V_1) > \mu_A(V_1) > \mu_C(V_1)$ and $\mu_C(V_6) > \mu_A(V_6) > \mu_B(V_6)$ In fact, there are all types of relationships between these measures that are induced by the system of payments.

Developing these systems (which appears to be quite like operating with sets) is at the center of measure theory.

Notation

In order to proceed, we will need additional notation to assist and support us. Typically, the symbol used in measure theory is $\mu(A)$, which means "the measure of the set A." For example, the notation $\mu(A_1 \cap A_2 \cap A_3)$ denotes the measure of the set, which is the intersection of the objects in the sets A_1, A_2 and A_3 It says nothing about how we actually take the measure but instead simply signals our intent to carry out the measure procedure.

We will use the *integral* sign the same way. Like the $\mu(\ )$ notation, $\int$ simply announces that we will be measuring a collection of objects. They may be discrete objects, intervals on the real line, or volumes of space (or combinations of all these different objects.) For example, if A is a collection of analyses in a clinical trial, then the notation $\int_A d\mu$ simply

---

* For this simple example, we are going to forego a formal validation that is indeed a measure.

means that we want to accumulate the measure or value of those analyses in set A using measure μ. Now we do not know what measure μ is at this point and will define it later. However, this is how we will use it. This can be a little disconcerting to an enthusiast of integral calculus.*

---

* Calculus students are used to a collection of formulas denoting how to integrate, such as $\int \cos(x)\,dx = \sin(x)$, or $\int_t^{\infty} \lambda e^{-\lambda x} dx = e^{-\lambda t}$. However, it is useful at this point to take a step back and see what we are doing. The classic way to view these standard integration rules is that we are accumulating "area under the curve," and of course, many times, that is not a wrong perspective. However, another approach is to say that we are taking the measure of a collection of points—in these circumstances, an interval on the real line. From this perspective, each of these formulas provides a different measure of the same interval. For example, consider an interval (a, b) on the positive real line. Then we know

$$\int_a^b dx = b - a : \int_a^b \cos(x)\,dx = \sin(b) - \sin(a) : \int_a^b \lambda e^{-\lambda x} dx = e^{-\lambda a} - e^{-\lambda b}.$$ Each of these three integrals does something different with the interval (a, b), i.e., each measures the (a, b) interval but uses a different measuring tool. For example, $\int_a^b dx = b - a$ denotes that the measure of an interval as simply its length. This is known most famously as Lebesgue measure. However, the other two definite intervals demonstrate that there are additional ways to measure the same interval, each providing a different answer. In fact, there are uncountably many measuring tools (many of which you already know) that provide the means to measure intervals of real numbers. Thus, when we are taking a definite integral, we are measuring the interval, and the integrand is the measuring tool.

From a measure theoretic perspective, there is no theoretical difference between measuring the real line by counting a subset of whole numbers on the one hand and completing a computation involving the length of the interval as the other. From the measure theory perspective, the only difference is the measuring tool.

Working with Measure's First Three Properties

Taking the measure of simple sets is straightforward. However, commonly, simple sets have little interest for us. We are interested instead in the measure of complex sets.

We will see that to find the measure of complex sets, we will build the complex set up from a collection of simple sets, using the set operators of union, intersection, and complement. If we are to find the measure of the complex set, we must identify the measure operator equivalents of these set functions' operations. We will begin to do this in this chapter.

Measure is typically taught as having four properties. This chapter will focus on measure's first three, leaving countable additivity to its own chapter. These first three properties are quite natural and intuitive, permitting us to develop several examples using the measure concept. With the experience we gain from these examples, we will be able to appreciate the need of the fourth property.

Review of the Sample Space and Sigma Algebras

Recall that the sample space (Ω) is the beginning source of set elements that interest us. The members (ω) of Ω are the building blocks of sets that hold the greatest interest. The set Ω can have a small number of events (for example, the number of patients in an infectology ward on a given day) or it can have an immense number of sets (the individual cubic nanometers of atmosphere over the Pacific Ocean).

The limitations of the constituents of Ω reside only within the scope of the problem and the imagination of the worker.

Once Ω is established as the foundation, the σ-algebra Σ is constructed. Think of Σ as a set generator; it is nothing more than the collection of sets built from a combination of the elements in Ω using the elementary set operations of unions, intersections, and complements.

Every element ω_i that is contained in Ω is also contained in Σ. Σ also contains the null set. In addition, Σ contains every possible union of different elements in Ω, first taken two at a time $\{\omega_1 \cup \omega_2\}$, $\{\omega_1 \cup \omega_3\}$, $\{\omega_1 \cup \omega_4\}$, ..., then three at a time, and so on. Next, Σ contains all the intersections, then unions of intersections, and then intersections of unions in all their complexity. From here, the process of building Σ continues, this time including complements of sets.

Thus, even when Ω is small, Σ can be quite large[*], and when Ω is large (e.g., the cubic nm of extravascular space), then the σ-algebra Σ can be quite overwhelming.

Measure versus Measurable Functions— Properties of Measure

Once Σ is identified, we are free to create a measurable set function on it. Remember that the only properties that a measurable function requires is that it must be nonnegative and that every value that it takes must map back to a set in Σ. Recall that we have tremendous freedom in defining measurable functions.

However, the actual measure of a set requires a more intricate operation than that conducted by a measurable function. Measure assigns not just a value, but content to a set. In order for a measurable function to be a measure, it must have the following four properties (three of which we focus on in this chapter):

> *Measure Property 1*
> If set A is a member of Σ, then $\mu(A)$ (called "the measure of A") must be a nonnegative real number.

---

[*] If, for example, Ω contains three and only three elements, Σ contains over thirty elements.

Measure Property 2
If μ is a measure on (Ω, Σ), then $\mu(\emptyset) = 0$.

Measure Property 3
If sets A and B are both elements of Σ such that A is contained in B, then $\mu(A) \leq \mu(B)$. Another way to say this is that if B contains A, then $\mu(B) \geq \mu(A)$.

Measure Property 4—Countable Additivity
(Discussed in the next chapter):
If the infinite sequence of disjoint sets A_n is contained in Σ, then $\mu\left(\bigcup_{n=1}^{\infty} A_n\right) = \sum_{n=1}^{\infty} \mu(A_n)$.

Of these properties, property 1 (nonnegative value) is the only one that measurable functions and measure have in common.

We can now explicate each of these three properties and their implications for measure.

Measure Property 1
If set A is a member of Σ, then $\mu(A)$ (called "the measure of A") must be a $\mu(A)$, nonnegative real number.

Just like measurable functions, a measure μ is a set function. This real number, $\mu(A)$, is the measure of, the content of, or the value of the set A And again, like a measurable function, its assigned value must be to a nonzero number. How the set is converted into a number is the property of the measure. However, the measure or content itself must be a nonnegative number.

Measure Property 2
If μ is a measure on (Ω, Σ), then $\mu(\emptyset) = 0$.

This statement buttresses the notion that the measure provides value or content to sets residing in Σ by permitting no value or content to the empty set. Even though the set $\emptyset$ resides within Σ, the measure we attach to it is, by definition, zero. For example, while one can quite reasonably define a measure based on the number of analyses in a clinical trial, it makes little sense to ask what is the measure or content of "no analysis." The statement $\mu(\varnothing) = 0$ is a mathematical statement of that reality, thereby making the concept of measure more universally practical.

However, the statement that $\mu(\varnothing) = 0$ has other important implications for us. For example, select two sets A and B from a (Ω, Σ) collection of sets such that the two sets are disjoint. Then we know that $A \cap B = \varnothing$, and therefore, by definition for any measure μ, $\mu(A \cap B) = \mu(\varnothing) = 0$.

Thus, the measure of the intersection of disjoint sets (however that measure may be defined) must be zero.

We will see that this simple implication, from among the most intuitive of measure theory principles, has critical implications for our work in applying measure theory to clinical research.

Measure Property 3

If sets A and B are both elements of (Ω, Σ) such that A is contained in B, then $\mu(A) \leq \mu(B)$. Another way to say this is that if B contains A, then $\mu(B) \geq \mu(A)$. This is known as the principle of monotonicity.

It might not be evident at first blush, but properties 2 and 3 tells us how to compute the measure of the union of sets.

Measure of the Union of Two Sets

This is our first real attempt to find the measure of a combination of sets. It is important to understand this concept completely because not only does it require us to review what we know about set theory, but it will also provide much new practice in building up the measure of complicated sets from simple ones.

For this development, let's assume that we are working from a standard (Ω, Σ) collection of sets and that sets A and B are selected from Σ

To begin with, we note from property 2 ($\mu(\emptyset) = 0$) that if our two sets are disjoint, then $A \cap B = \emptyset$, and thus $\mu(A \cap B) = 0$.

So the solution for the measure of the intersection of disjoint sets is already available to us. However, what is the measure of their union?

In order to examine this concept using a simple illustration, consider a set A_1 that contains a single element ω_1. Then $\mu(A_1) = \mu(\omega_1)$.

Now consider a set A_2 with two elements $\{\omega_1, \omega_2\}$. What is $\mu(A_2)$? If ω_2 is the same element as ω_1, then

$$\mu(A_2) = \mu(\{\omega_1, \omega_2\}) = \mu(\{\omega_1, \omega_1\}) = \mu(\{\omega_1\}) = \mu(A_1) = \mu(\omega_1)$$

However, if the element ω_2 is not the same as ω_1, (which means that these elements are disjoint) and if measure is to serve as a process of accumulation, then the natural conclusion is that $\mu(A_2)$ is the sum of the $\mu(\omega_1)$ and $\mu(\omega_2)$. This is how we would expect the accumulation process to work.

Now let's talk in general about two sets A and B. Another way to write the set A is $A \cup \Omega$. Still another way to write Ω is $\Omega = B \cup B^c$. Then we have

$$\begin{aligned} A &= A \cap \Omega \\ &= A \cap (B \cup B^c) \\ &= (A \cap B) \cup (A \cap B^c) \end{aligned}$$

Analogously, the set B can be written as $B = (B \cap A) \cup (B \cap A^c)$.

Thus

$$\begin{aligned} A \cup B &= (A \cap B) \cup (A \cap B^c) \cup (B \cap A) \cup (B \cap A^c) \\ &= (A \cap B^c) \cup (B \cap A^c) \cup (A \cap B). \end{aligned}$$

This we have seen from the chapter on set theory. Figure 1 is reproduced here.

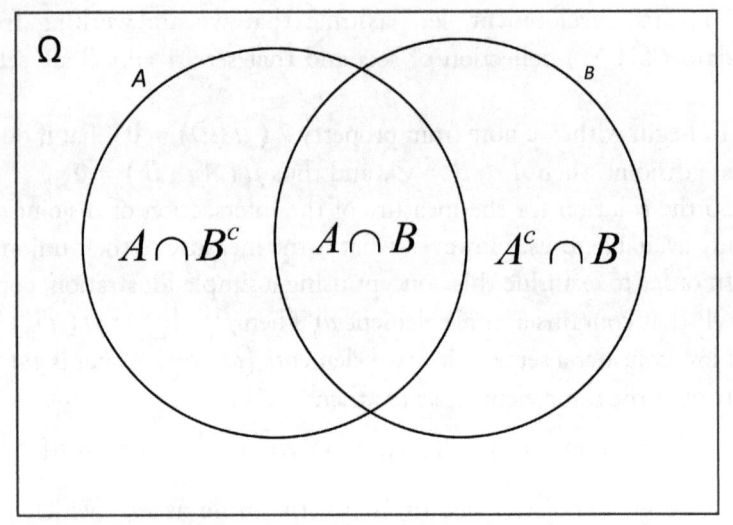

Figure 1. Demonstration of unions, intersections and complements using 2 nondisjoint sets A and B.

From this formulation, we can see that the set $A \cup B$ is composed of three subsets: (1) the part of A that does not contain B, (2) the part of B that does not contain A, and (3) the elements common to both A and B, namely $A \cup B$.

By this restructuring of $A \cup B$ into the union of three sets, we notice that the terms on the right are pair-wise disjoint. For example, for an element to be in $A \cap B^c$, it must be in B^c, which excludes it from the sets $A^c \cap B$ and $A \cap B$. Since these sets are disjoint, we can sum their measures.

$$A \cup B = (A \cap B^c) \cup (B \cap A^c) \cup (A \cap B).$$
$$\mu(A \cup B) = \mu(A \cup B^c) + \mu(A^c \cup B) + \mu(A \cup B).$$

This is the most general solution for the measure of the union of two sets and one that we will take advantage of.

Special Cases of the Union of Two Sets

With this as background, we can now examine some special cases of $A \cup B$. For example, if $A = B$, then our intuition tells us that $\mu(A) = \mu(B)$. We can show that from our previous formulation of $A \cup B$ as

$$\mu(A\cup B) = \mu(A\cap B^c) + \mu(A^c \cap B) + \mu(A\cap B).$$

In the case where $A = B$, then
$A\cap B^c = \emptyset$, $A^c \cap B = \emptyset$, and $A \cap B = A = B$, so $\mu(A\cup B) = \mu(A) = \mu(B)$.

As another example, if A and B are disjoint, then $\mu(A\cap B) = \mu(\emptyset) = 0$. We can then write $\mu(A\cup B) = \mu(A\cap B^c) + \mu(A^c \cap B)$. Then writing $A \cap B^c = A$ and $A^c \cap B = B$ we have
$$\mu(A\cup B) = \mu(A) + \mu(B)$$

These have been the simple cases. But what if the two sets are neither equal nor pairwise disjoint, i.e., $A \neq B$ and $A \cap B \neq \emptyset$?

Our background in set theory will help us find and bound $\mu(A \cup B)$. We know that we can write

$$A = (A\cap B)\cup(A\cap B^c)$$
$$B = (A\cap B)\cup(A^c \cap B)$$

so that

$$\mu(A) = \mu(A\cap B^c) + \mu(A\cap B)$$
$$\mu(B) = \mu(A^c \cap B) + \mu(A\cap B).$$

We sum these two expressions to find that

$$\mu(A) + \mu(B) = \mu(A\cap B^c) + \mu(A\cap B)$$
$$+ \mu(A^c \cap B) + \mu(A\cap B)$$
$$= \mu(A\cap B^c) + 2\mu(A\cap B) + \mu(A^c \cap B)$$

However, we know from before that

$$\mu(A\cup B) = \mu(A\cap B^c) + \mu(A^c \cap B) + \mu(A\cap B)$$

which, comparing term by term, is less than

$$\mu(A\cap B^c)+2\mu(A\cap B)+\mu(A^c\cap B).$$

We may write in general that

$$\mu(A\cup B)=\mu(A)+\mu(B)-\mu(A\cap B)$$
$$\mu(A\cup B)\le\mu(A)+\mu(B).$$

Using these same three simple properties of measure, we may find the measure of complements of sets. We begin by writing $\Omega = A \cup A^c$. Since the sets A and A^c are mutually exclusive, we can write $\mu(\Omega) = \mu(A) + \mu(A^c)$, or $\mu(A^c) = \mu(\Omega) - \mu(A)$. If the measure of the sample space Ω is finite and known, then we can find the measure of A^c from $\mu(A)$.*

As another example, let's use what we know about manipulating sets and their measure to show that if $A \subset B$, then $\mu(A) \le \mu(B)$. Again, our intuition tells us that this should be true; if set A is contained in set B, then B contains A plus something else. If the measure of that "something else" is not zero, then $\mu(A) \le \mu(B)$. With this helpful thought process behind us, let's now apply what we know of measure theory to this simple problem.

We know that

$$\mu(A\cup B)=\mu(A\cap B^c)+\mu(A^c\cap B)+\mu(A\cap B).$$

In this case where $A \subset B$, then $A \cup B = B$, $A \cap B^c = \emptyset$, and $A \cap B = A$.

Thus, $\mu(B)=\mu(A^c\cap B)+\mu(A)$. If $\mu(A)\ne\emptyset$, then $\mu(A^c\cap B)\ge 0$, and $\mu(B)\ge\mu(A)$. If $A=\emptyset$, then this equality reduces to

$\mu(B) = \mu(\Omega \cap B) + \mu(\emptyset) = \mu(B)$. This is a demonstration of measure property 3.

---

* If $A = \emptyset$, then $A = \Omega$, and $\mu(A^c) = \mu(\Omega)$.

Returning to $\mu(B) = \mu(A^c \cap B) + \mu(A)$, simple subtraction reveals that $\mu(A^c \cap B) = \mu(B) - \mu(A)$, another finding of which we will make use.

Summary

So in this chapter, we have explored three properties of measure. Measure is based on sets; we will always apply measure to sets. Thus, our ability to manipulate measure is tied directly to our ability to manipulate sets. Exploring some of the implications of the first three properties of measure permitted us to develop the measure of the intersection of sets and the measure of the union of sets.

In fact, from three simple properties of measure, we can find the measure of unions and complements of sets. This is one of the most important components of measure theory that we will use. First, we will construct an (Ω, Σ) collection of sets. Then we will identify the set whose measure we want by building that set up from an intelligent combination of set operations of unions and intersections.

This combination of set operations will be paralleled by adding and subtracting the measure of sets that will get us to the measure of the set we ultimately desire.

We are now ready to examine the fourth property of measure—countable additivity.

Property 4 of Measure: Countability

Remember that the first three properties of measure (measure must be nonnegative, measure of the null set is zero, and the measure of one set that contains another) allow us to assemble the measure of sets that are built up from other sets.

The motivation for properties 1–3 (nonnegativity, the zero value of emptiness, and the relative value of sets containing one another) comes from the need and desire to bring a useful and intuitive concept to measure. We want it to assess the content or value of an item in a way that matches our intuition (e.g., the value is never negative, and the value of "nothing" is zero), and we also want measure to accumulate (and not decrease) over a collection of sets, each one containing the preceding one. These three properties built on a basis of set theory ensure that.

However, property 4 (countable additivity) has a different motivation. It has little to do with how we assign measure to a set but is more focused on the actual utility of measure.

Specifically, given that measure is assigned to a collection of sets, how can it be used to assign measure to other more complex sets that are formed from that original collection?*

For example, suppose that we want to measure a set A. We know that A can be produced from a collection of sets $\{A_i\}$. and that each have known measure. If this collection of sets whose measure we know builds the set A from the union and intersections of the members of $\{A_i\}$, then we can build up to the measure of A by what we know of computing the measure of unions and intersections of $\{A_i\}$, no matter how many sets

---

* We have seen a rudimentary example of this with the determination of the measure of the union of two sets.

are contained in $\{A_i\}$. This is one way in which countable additivity is important.

Measure Property 4—Countable Additivity
If the infinite sequence of disjoint sets A_n is contained in Σ, then $\mu\left(\bigcup_{n=1}^{\infty} A_n\right) = \sum_{n=1}^{\infty} \mu(A_n)$.

This can be proved using an induction argument.*

Note that the upper bound of the index in this property is infinity. There is another concept where the upper bound is finite, termed finite additivity, i.e., $\mu\left(\bigcup_{i=1}^{n} A_i\right) = \sum_{i=1}^{n} \mu(A_i)$. This is a consequence of the countable additivity property, which can be easily demonstrated.†

Nonnull intersection sets within the sequence of sets $\{A_i\}$ require us to modify the assertion of property 4 to say that if the sets are not disjoint, then $\mu\left(\bigcup_{i=1}^{\infty} A_i\right) \leq \sum_{i=1}^{n} \mu(A_i)$. ‡

---

* The veracity of this property can be developed through an induction argument. For $k = 1$, $\mu(A_1) = \mu(A_1)$. If we assume

$$\mu\left(\bigcup_{n=1}^{k} A_n\right) = \sum_{n=1}^{k} \mu(A_n),$$ then for the $k+1^{st}$ set, A_{k+1} is disjoint from $A_1, A_2, A_3, \ldots A_k$ and is therefore disjoint from their union. Thus,

$$\mu\left(\bigcup_{n=1}^{k+1} A_k\right) = \mu\left(\bigcup_{n=1}^{k} A_k \cup A_{k+1}\right) = \mu\left(\bigcup_{n=1}^{k} A_k\right) + \mu(A_{k+1})$$

$$= \sum_{n=1}^{k} \mu(A_n) + \mu(A_{k+1}) = \sum_{n=1}^{k+1} \mu(A_n),$$

completing the induction argument.

† We note that $\mu\left(\bigcup_{n=1}^{\infty} A_n\right) = \sum_{n=1}^{\infty} \mu(A_n)$. Now let's choose our collection of sets such that for $n = 1$ to k $A_n \neq \emptyset$. However, for all $n > k$, $A_n = \emptyset$. Then

$$\mu\left(\bigcup_{n=1}^{\infty} A_n\right) = \mu\left(\bigcup_{n=1}^{k} A_n \cup \bigcup_{n=k+1}^{\infty} A_n\right) = \mu\left(\bigcup_{n=1}^{k} A_n \cup \bigcup_{n=k+1}^{\infty} \emptyset\right) = \mu\left(\bigcup_{n=1}^{k} A_n\right) = \sum_{n=1}^{k} \mu(A_n).$$ Thus, for this collection of A_n, the infinite union reduces to a finite union of exactly the A_n sets that we want.

‡ We have seen this demonstrated in the previous chapter with the union of two sets.

This fourth countable additivity property of measure, while appearing somewhat abstract right now, is actually quite important. It permits us to deconstruct the measure of a large union of sets into the measures of the individual constituents of these sets. In addition, if the individual sets that compose the union are disjoint, we can simply sum their measures for equality. Much of the developmental work of measure theory is based on the ability to deconstruct the union of sets into an equivalent union of different but disjoint events and then use the property of countable additivity to sum the measure of these disjoint sets.

With this as background, we are now ready to consider the content of a clinical research analyses, in a way that we can apply a measure theoretic approach to its accumulation.

An Interlude ...

Let's just pause for a second and see where we are.

The preceding set and measure theoretic preamble permits us to compute the measure of combinations of sets. If we start with a collection of sets, each of whose measure we know, we can now compute the measure of a more difficult and intricate set by using the rules of measure in parallel with the set operations (unions, intersections, and complements of the original, simpler sets) to build up the measure of the final set.

But what does this have to do with clinical research?

We began with the notion of duality, i.e., the idea that a single estimate from clinical research can be evidence both for benefit and for harm. There we described a process by which a region of plausible values of effect sizes (i.e., a plausible interval) could be parsed into one for benefit and the other for harm. For benefit, this is $\chi_i^{(b)}$ and its benefit function is $\mathbf{Y}_b\left(\chi_i^{(b)}\right)$.

Our concept is that this region and function can be identified for each analysis. However, we need to accumulate them over all analyses, $\int_{A \subset \Omega} \mathbf{Y}_b\left(\chi_i^{(b)}\right)$. But how do we compute this when the individual analyses overlap? What do we do about the redundancy of observations and variables that are used repeatedly in succeeding analyses?

Specifically, the collections of analyses utilize overlapping collections of observations or variables. An examination of a mean difference and the assessment of an effect using a general linear model with adjustments for covariates are different analyses but can have observations in the first analysis also included in the second analysis and variables in the second analysis containing variables in the first analysis. And the more analyses that were conducted, the more intense the analysis is likely to be (figure 1).

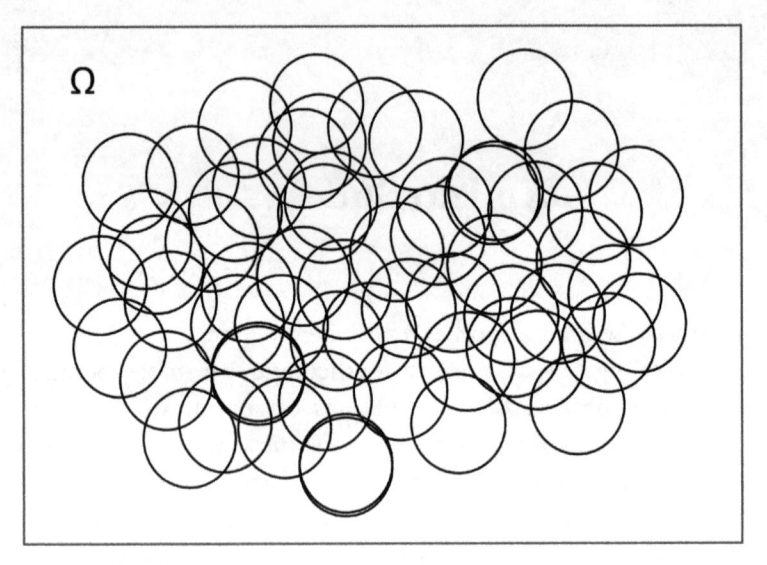

Figure 1. Overlap of observations and variables utilized in analyses in a clinical trial.

If we are, in some fashion, to integrate or accumulate the plausible intervals of benefit and harm over all relevant analyses, how can we conduct this integration with this intense degree of redundancy?

This is where we use our background in set and measure theory. From this perspective, we now know how to create and manage this complex situation.

Specifically, we create a universe containing all these analyses from a clinical research effort and call that universe Ω. We then create a σ-algebra of these analyses, Σ which consists of all the subsets of Ω.

We know that the sets described in the previous paragraph are members of the σ-algebra because each analysis is a member of Ω and the sets are constructed from our standard set operations of unions, intersections, and complements. In fact, a set (e.g., that of the primary analyses) is but one set of a large collection of analyses that can be assembled and combined in any way that we like.

Of course, there are certainly rules that we will use to direct our attention; we will allow the principles of epidemiology and our intellectual discipline to focus on some sets of analyses while dismissing others, but theoretically, each of these analysis sets is available for inspection. However, we need a *measure* for each analysis, one that deals sensibly with the redundancy concern.

However, if we can establish such a measure (it is what we soon call our psi measure, ψ – measure), then an entirely different vista opens before us. We will be able not just to measure any analysis but also measure any set of analyses in Σ. Such an identification permits us, for example, to discount analyses based on their relatively small measure. We can compute the content of the primary analyses in a clinical research effort or the content of all prospectively declared analyses.

It would allow us to measure "exploratory" analyses, comparing and contrasting their measure to that of prospective analyses. And the large size of Σ can produce many interesting, heretofore-unconsidered analysis sets where content can be assessed.

Let's assume for a moment that we had a measure μ of each analysis A_i, $i = 1, 2, 3, ..., n$ contained in Ω that was based on the number of observations and variables used in each analysis, and we wish to compute $\psi(\Omega) = \psi\left(\bigcup_{i=1}^{n} A_i\right)$. Figure 1 tells us that this would be a complicated operation due to the degree to which the analyses overlap, but we know how to proceed.

In addition, we can follow different paths of analyses to build up this union. For example, we might start in figure 1's upper right corner or its lower left corner or even its center. Out intuition tells us that the solution to $\psi\left(\bigcup_{i=1}^{n} A_i\right)$ will be the same regardless of which path we take, but the contribution of any particular analysis to that union will be path-dependent due to the intense overlapping of the analyses. This is an observation of which we will take advantage.

So with our goal in mind and set and measure theoretic background and context in place, we are ready to define an analysis measure and examine its implications.

Functions and Measures
on Analysis Regions

We are now almost ready to develop a quantity that, for the moment, we will call the content of an analysis.* But before we embark on this, we should review our assumptions.

This critical review requires that we understand the properties of the elements of Ω and Σ. Since we know that any function or measure that we develop based on (Ω, Σ) must be measurable, we should ensure that each element of Ω is imbued with the properties that our functions f and measures μ will inspect and recognize. This is critical because we need to make sure that our assessment of analyses through a function or a measure be measurable, i.e., the elements of Ω have the property that our content can assess.

In addition, a thorough understanding of Ω also provides a sound foundation for us to develop, manipulate, and ultimately deploy set functions and measures with confidence.

What Constitutes an Analysis?

A major product of a clinical research effort is its collection of analyses. Let's denote the ith analysis of such an effort as ω_i (order does not matter at this point). Then we will define Ω as the congregation of

---

* Please note that I am staying away from the use of the term *information*. While that term might be natural and intuitive, I want to avoid any confusion with the well-developed science described by the term *statistical information theory*, which is related to coding theory, channel theory, and entropy.

all these analyses from the study, i.e., the superset $\{\omega_1, \omega_2, \omega_3, ...\omega_n\}$ or $\{\omega_i, i = 1,..., n\}$.

This we have stated before, but what exactly does this mean? What is an analysis, and what are its properties?

An analysis is complicated. While it is easy to think of an analysis as a result (i.e., a hazard ratio), an analysis is a basket of properties associated with a computation. It has many constituent parts.

For example, the analysis has to be designed. What question is the analysis designed to answer? Is the analysis to be prospectively designed or retrospective? On what instrumentation will it be based (e.g., if the analysis was of heart structure, then was that measurement based on magnetic resonance imaging or echocardiography with contrast?).

The analysis is also based on a specific collection of observations. It uses certain specific variables in the data set. It utilizes a specific formula classifying it as a type of analysis, e.g., regression analysis, nonparametric analysis, survival analysis, etc.). We can add other characteristics (is the analysis a subgroup analysis?). It generates an estimate of effect. It also produces a standard error, has a known precision, and can be subject to bias.

Thus, although it is easy to simply say that each analysis is a member of the space, i.e., $\omega_i \subset \Omega$, there are many characteristics or analysis properties contained in each ω_i. And each basket of properties is quite rich, and we can enumerate them as the properties of ω_i, $\delta_i(j), j = 1, 2, 3...,$.

While this may be a new perspective on an analysis, it is not a new concept from our set-theoretic perspective. When we assemble, for example, a collection of patients in a clinical trial—say, $p_1, p_2, p_3,...p_n$ —and create a space Ω for them, we are inferentially gathering individual traits of each patient (demographics, phenotypes, comorbidity, therapy assignment, compliance, etc.). Each individual is a collector of many different traits of that individual. We are simply applying this familiar concept to the set of analyses.*

This understanding gives us facility in working with the contents of our analysis space Ω. For example, one such property of an analysis is the question that motivated the analysis. Denote this property of ω_i as q_i. Now this question q_i can be broad (e.g., What is the effect of the exposure being studies on the overall health of the exposed population

---

*   In some circles, this basket of items could be considered metadata.

as compared to the unexposed population?) or narrow (e.g., What is the effect of a single dose of the exposure on the change in blood pressure over time?).

Since broader questions q can contain more specific questions, it can be anticipated that there will be an interest in focusing on the analysis content over subsets of analyses that address important components of the relevant question. Such a subset can be denoted as $\{\omega_i / q_i \subset q\}$ or $\omega_i \subset A_q = \{\omega_i / q_i \quad q\}$, which means that the analysis is a member of the set of analyses responsive to question q Since q_i is a property of the analysis ω_j, we can aggregate these analyses, the aggregation being a subset of Ω, which resides in Σ And since they have the properties that we are interested in, they will be measurable and be available for content assessment.

Regions of Analyses

A region of analysis is simply a collection of analyses that has a common characteristic or property.

All the analyses that provide an answer to a question q constitute a region of analyses. All subgroup analyses comprise another region. The collection of mixed models assessing gender-mediated effects on therapy for systolic blood pressure is another. Our broad definition of ω_i offers a wide latitude in creating and ultimately evaluating collections of analyses. Since it contains so many analysis properties, analyses can be assessed using a wide range of measurable functions and measures.

Investigators have broad authority in conducting clinical trial analyses. For example, suppose that the prompting question of the research effort is "Does the provision of allogeneic cells into a subject's heart reverse the progression of heart failure?" The investigator can choose from many candidate variables (e.g., mortality, exercise tolerance, quality of life, biomarkers). They can also conduct evaluations on different cohorts (e.g., only females or only patients greater than sixty years of age), as well as implement different but related estimates of the effect of therapy (e.g., nonparametric tests, survival analyses, general linear model assessments). Each of these is a region of analysis.

So our purpose will be to accumulate analysis content over these regions of analyses.

But in addition, we can provide weight to these analyses. The accumulation process over the rich properties of ω_i permit us not to just assess the content of ω_i in multiple dimensions at once. Taking advantage of these analysis features permits the circumstances and characteristics of the analysis to appropriately affect its contribution to the evidence addressing the question at hand.

For example, if analysis ω_i is responsive to question q, i.e., $\omega_i \subset A_q = \{\omega_i / q_i \quad q\}$ but is exploratory and the investigators believe that exploratory analyses are not considered contributory to the answer to question q, then that analysis's contribution to answering question q can be set to zero. This is easily accomplished by having the Σ measurable function $f(\omega_i)$ that is integrated over the content of ω_i to be zero if ω_i is exploratory. We know $f(\omega_i)$ is measurable since the exploratory characteristic is embedded within ω_i.

As another example, we can simultaneously assess the benefit estimated by ω_i and also the content of ω_i. The benefit we have discussed earlier; we now know that it is a measurable function on (Ω, Σ) because it can be constructed from traits of ω_i.

We can now find that measure, beginning with the concept of the content of an analysis ω_i.

Defining the Content of an Analysis

We have discussed how each analysis ω_i is a basket of descriptors that describes and summarizes the analysis. Focusing on two of its items will generate a measure for us.

Let's begin with any one of the analyses $\omega_i \subset \Omega$ in a clinical research effort. The first item we will use is the collection of subjects who contributed data to the analyses. This is not the number of subjects but the subjects themselves. Each subject's identity is typically denoted as a depersonalized depiction or ID. It is a unique pointer to a single individual in the analysis.

In order to describe this, we can create a vector or a collection of ID numbers denoting all the individuals included in the analysis. We will call this collection $\underline{\mathbf{n}}_i$ (the subscript i links it to analysis ω_i). We will also denote the number of individuals this represents* as n_i.

We would expect that in two different analyses in the same clinical trial, there would be substantial redundancy in the subjects used for each one. In fact, the collection of individuals for the first would sometimes be the same as the collection for those in the second. In this case, $\underline{\mathbf{n}}_1 = \underline{\mathbf{n}}_2$ and, of course, $n_1 = n_2$. Alternatively, if several subjects who took part in analysis 1 did not contribute to analysis 2, then $\mathbf{n}_1 \neq \mathbf{n}_2$, and $n_1 > n_2$.

We can follow the same procedure for working with the variables that are part of an analysis. Let's define $\mathbf{v}_i$ as the collection of variables that are evaluated for the n_i subjects in analysis ω_i. We will also define the total number of variables utilized in analysis ω_i is v_i.

---

* If one thinks of the collection of ID numbers as a vector $\underline{\mathbf{n}}_i$, then n_i is the dimensionality or the tuple of that vector.

93

As an example, consider a clinical trial that has randomized 75 subjects to each of a control group and a treatment group. The purpose of the analysis is to address the prospectively asked question "What is the effect of therapy on the difference in change in mean diastolic blood pressure between the two groups?"

In this circumstance, the vector $\mathbf{n}_1$ contains 75 entries, each one being the ID of an individual whose data (variables) were used in the analysis and $n_1 = 75$. Since three variables are involved (baseline and follow-up DBP, plus the variable denoting the therapy group), then $\mathbf{v}_1$ contains the three variables names (not the data points themselves) and $v_1 = 3$.

The Content of an Analysis

Using what has become our standard definition for ω_i, which is an analysis contained in Σ, the σ-algebra of Ω, we define the content of analysis ω_i as $\psi(\omega_i)$ and write $\psi(\omega_i)$ as $\psi(\omega_i) = n_i v_i$.

The content of an analysis is quite simple—simply the product of the number of participants whose data is included in the analysis and the number of variables that are required for the analysis. Notably, it does not include any of the other components of ω_i, e.g., the question that generated the analysis, the design characteristics of the analysis, or the effect size produced by the analysis. Instead, the definition of analysis content is based solely on the data that contributed to the analysis. The unit of ψ is simply subject-variables.

At this point, we are not in a position to claim that $\psi(\omega_i)$, defined as $\psi(\omega_i) = n_i v_i$, is a measure. We will have to examine if it meets all of our four-measure criteria. Until we do, we will simply call ψ the content of the analysis.

However, we can explore this concept and appreciate at least some of its implications. As an example, consider a clinical trial in which the analysis being conducted is the comparison of the difference in the change in DBP from baseline to six months between the treatment group and the control group.

Eighty subjects contribute to the analysis, and three variables were required (baseline DBP, follow-up DBP, and the treatment group identifier). In this case, $\psi(\omega_i) = (80)(30) = 240$. An analysis that evaluates 298 subjects for the effect of therapy on the change in systolic blood pressure (SBP) between two subgroup strata has content $(298)(1+3) = 1192.$*

From this simple formulation in this discussion, we can make the following observations about ψ-content:

1. Every analysis conducted has a content.
2. If the number of variables is fixed, then the larger the set of subjects used in an analysis, the greater the content of the analysis.
3. If the number of participants is fixed, then the greater the number of variables in the analysis, the greater the analysis content.
4. Analysis content is independent of the design features of an analysis.

Since ψ-content is based on simply the number of observations and the number of variables that are included in the analysis, items 1, 2, and 3 are self-evident.

The fourth observation above, however, requires attention. The content of an analysis is separate and apart from an assessment determining that analysis's probative value. Analyses of little value (e.g., an analysis conducted in a clinical trial that is irrelevant to the question under consideration) can have high content. Similarly, analyses that have great, strategic value may have relatively little content. While we will specifically deal with the concept of analysis value in a later chapter, it is clear that other features of the analysis that gauge the analysis's worth must also be integrated into this process.

As developed here, analysis content is separate and apart from analysis contribution. This purpose of ψ-content is to provide a mathematical basis for accumulating overlapping evaluations over their regions of analyses.

However, in order to achieve this goal, we must address the issue of overlapping analyses, the subject of the next chapter.

---

* The four variables are (1) baseline SBP, (2) follow-up SBP), (3) therapy assignment, and (4) the one variable on which the data are stratified. Variables—e.g., the number of strata divisions or the function of a variable (e.g., squares or logs)—are derivative variables, i.e., their content derives from other variables in the data set.

Analysis Redundancy

There is no question but that there is commonly redundancy between different analyses. Thus far, we have not addressed this critical concern.

For example, consider the content of an analysis ω_1 consisting of a general linear model that uses 50 subjects and 2 explainer variables for a dependent variable. From the previous chapter, $\psi(\omega_1) = (50)(20) = 100$.

Now consider a second analysis ω_2, which is a general linear model that studies the same 50 subjects and the same 2 explainer variables for the same dependent variable and, in addition, contains one more explainer variable. We compute $\psi(\omega_2) = (50)(3) = 150$. Yet even though the content of the second analysis is greater than that of the first, they both use the same subjects and also have two of three variables in common. This considerable overlap suggests that in the process of content accumulation, the content of the second analyses should be reduced or moderated if the first analysis's contribution has already been considered.

We begin our examination of this by recalling that the content of an analysis ω_i is $\psi(\omega_i) = n_i v_i$, where n_i is the number of subjects and v_i is the number of variables. Note that the content does not depend on the identify of these subjects—only the number of them. The same is true for the variable component of the analysis's content.

This will not be true for managing the degree to which two analyses overlap.

From our set theory perspective, the caliper of analysis redundancy between two analyses is simply the degree to which two sets (on which those analyses are based) are not disjoint. Thus, the overlap between two analyses is addressed by considering the intersection of these analyses

ω_i and ω_j, $\omega_i \cap \omega_j$. Let's denote the content of this intersection as $\psi(\omega_i \cap \omega_j)$ and define

$$\psi(\omega_i \cap \omega_j) = n_{ij} v_{ij}.$$

Here, n_{ij} and v_{ij} are the number of subjects and number of variables common to both analyses ω_i and ω_j. Assessing the commonality requires us to focus on not just the number of subjects and variables used in the two analyses but also the degree to which they are the same.

For example, consider two analyses in a clinical trial that has randomized 305 subjects. The first analysis incorporates 298 of these subjects and utilizes 5 variables. The second analysis incorporates 245 of these 298 subjects and utilizes 8 variables, 3 of which are common to the first analysis. Then we may compute

$$\psi(\omega_1) = n_1 v_1 = (298)(5) = 1490$$
$$\psi(\omega_2) = n_2 v_2 = (245)(8) = 1960$$
$$\psi(\omega_1 \cap \omega_2) = n_{12} v_{12} = (245)(3) = 735.$$

We define, in general, the content of the intersection of k analyses $\omega_1, \omega_2, \omega_3, \ldots \omega_k$ as

$$\psi\left(\bigcap_{i=1}^{k} \omega_i\right) = n_{i\ldots k} v_{i\ldots k}$$

where $n_{i\ldots k}$ is the number of observations common to all k analyses and $v_{i\ldots k}$ is the number of variables common to all k analyses.

Computing the Content of Analysis Unions

The ability to calculate the content of the intersection of analyses is precisely what we need to compute the content of analyses' unions, which is our goal. For example, from out previous measure theory development, we know that we can write the content of the union of analyses ω_i and ω_j as $\psi(\omega_i \cup \omega_j) = \psi(\omega_i) + \psi(\omega_j) - \psi(\omega_i \cap \omega_j)$. In this situation of clinical research program, we can write that the content of the union of two analyses as $\psi(\omega_i \cup \omega_j) = n_i v_i + n_j v_j - n_{ij} v_{ij}$. In the above example,

$$\psi(\omega_i \cup \omega_j) = \psi(\omega_i) + \psi(\omega_j) - \psi(\omega_i \cap \omega_j)$$
$$= n_i v_i + n_j v_j - n_{ij} v_{ij}$$
$$= 1490 + 1960 - 735$$
$$= 2715.$$

However, this relatively simple formulation has important consequences that we must now consider, understand, and accept if we are to work with it.

1. *Analyses have no common content if they have no common subjects.* If there are no common subjects between two analyses, ω_i and ω_j, then $n_{ij} = 0$,. The analyses are disjoint, and the intersection of the two analyses has zero content. This makes intuitive sense because in clinical research, subjects are assumed to operate independently of one another, making separate contributions to analyses. Thus, separate and disjoint collections of participants produce separate content for each analysis, but with no joint contribution, there is no joint content. Thus, the analyses are disjoint, and the content of their intersection is zero.
2. *If two analyses have no common variables, although $v_{ij} = 0$, the content of the intersection $\psi(\omega_i \cap \omega_j)$ will frequently be but need not be zero.* At first blush, in the case where $v_{ij} = 0$, using the reasoning of the previous observation just stipulated, it follows that the content of the intersection of the two analyses should be zero.

However, there is an exception that is well recognized in clinical research that we must incorporate. Variables—unlike participants—can be and commonly are interrelated, and the degree to which they are related to one another impacts the commonality of the analyses under consideration. For a fixed number of participants, if the collection of variables in analysis 1 is related to the collection of variables in analysis 2, then even though the variables are not explicitly common, their interrelatedness would convey a nondisjoint interaction. This complicating situation is explicitly managed and incorporated in a later chapter. At this early point in the development, we will simply say that in the absence of interrelationships between the variables, the absence of common variables between two analyses implies that the content of the interaction of the two analyses is zero.

3. *The content of the union of analyses is the sum of the analyses' content when the analyses do not overlap.* This is not unexpected at all and simply follows from our application of set and measure theory to the circumstance of accumulating the measure of disjoint sets.
4. *Analyses with substantial content can have different purposes and value.* This is a critical observation. The input to the content of two analyses' joint content is strictly mechanical. Therefore, two analyses can have substantial intersection content although from an intellectual or design perspective, they have different motivations and come to conclusions in different noetic dimensions.

Consider, for example, a clinical trial providing two analyses each involving the same 100 subjects. The first is an assessment of the impact of the randomly allocated therapy on mortality, the primary analysis of the study. The second is an exploratory assessment of the effect of therapy on a newly discovered cell phenotype using those same 100 subjects. Because the analyses use the same 100 subjects as well as the treatment assignment variable, there is considerable content redundancy. However, these two analyses have very different purposes and bring different value levels to the overall investigation. Thus, while there may be substantial redundancy in analysis content, the analyses may nevertheless play very different roles in the research enterprise.

Similarly, a large value of $\psi(\omega_i \cap \omega_j)$ does not suggest that the two analyses ω_i and ω_j address the same research question—only that they have common substrate (i.e., each draws from the same participants and the same variables).

5. *An analysis's content can be separate and apart from that analysis's value.* This again speaks to the difference between intellectual value and ψ-content. A survival model studying 453 subjects using three variables (time to event, censoring mechanism, and therapy assignment) has less content than a mixed model regression analysis on the same subjects that incorporates five different covariates. However, the importance of the mortality evaluation is greater than the regression analysis if the effect of therapy on the death rate in the study was the most important question. The content of an analysis is simply based on the number of observations and number of variables used in the evaluation and is quite rote. However, an analysis's value depends on other properties of ω_i, $\delta_i(j)$, $j = 1, 2, 3...$, e.g., the interrogation that the analysis addresses and whether that analysis was prospectively declared. The content of an analysis is separate and apart from its purpose and value.

Given observations 4 and 5 above, why should we bother developing a mechanical content that has no cerebral input and is devoid of independent and of critical considerations of epidemiology, research design issues, or the investigator-determined priority of analyses? Aren't these latter concerns the really interesting and necessary metrics? Why not build a construct based on them?

The answer is yes, those advanced concepts are essential to drawing conclusions from research efforts. However, at this point, we are not yet ready to mathematically include them (this occurs later in this book).

We are constructing cornerstones now. Like any foundation, it must be objectively assessable, reproducible, and durable. While intellectual assessments of acceptable analyses change over time*, our foundation must be fixed. The mechanical ψ-content serves handsomely in this regard since it is incontrovertible that observations and variables are required for analyses.

---

* For example, in the 1970s, subgroup evaluations and exploratory analyses were as admissible as primary outcome assessments because they are each derived from clinical trials.

There are, of course, other metrics (quality of research design, prospective declaration, etc.) that must and will play an essential role in assessing the quality and value of the analysis, but they will come later.* The absence of their contribution at this point is why we use the neutral term *content* to describe ψ.

But the question now before us is whether ψ-content is really a measure at all. This is the topic of the next chapter.

---

* In measure-theoretic language, ψ will ultimately be the measure, while the important epidemiologic and intellectual contributions will be measurable functions that operate on the (Ω, Σ) measure space.

Converting ψ-Content to ψ-Measure

This chapter, by necessity, is a fairly technical chapter. For those who want to be convinced that ψ-content is indeed a measure, please read through the material here in detail. However, if you are comfortable assuming that this mathematical development is correct, then feel free to skip to the chapter summary.

To recap, we have developed a set function that converts the element of a clinical trial analysis into something we have defined as content. We wanted this content to key off what was contained in an analysis ω_i to ensure that our content was measurable, but having done that, we had tremendous freedom in the selection of the elements of the analysis ω_i that would be incorporated in our content development. We chose $\psi(\omega_i) = n_i v_i$ and $\psi(\omega_i \cap \omega_j) = n_{ij} v_{ij}$. This was not the only choice of a definition for content (other possible selections are discussed in the later chapter on limitations); but it is a simple one, and we have explored this structure's properties and weaknesses.

Our next task is to demonstrate that ψ is a measure. Since this is an important step, we will need to develop these arguments formally.

Let Ω as the congregation of all analyses conducted in the clinical trial, i.e., the superset $\{\omega_1, \omega_2, \omega_3, ...\omega_n\}$ or $\{\omega_i, i = 1,...,n\}$. Denote Σ its σ-algebra. Remember that the four assertions required to demonstrate ψ is a measure are the following:

> Assertion 1: For $\omega_i \subset \Sigma$, $\psi(\omega_i) \geq 0$.
> Assertion 2: $\psi(\varnothing) = 0$.
> Assertion 3: If $\omega_i \subset \omega_j$, then $\psi(\omega_i) \leq \psi(\omega_j)$.

Assertion 4: $\psi\left(\bigcup_{i=1}^{\infty} \omega_i\right) \leq \sum_{i=1}^{\infty} \psi(\omega_j)$.

We will see that three of these assertions are quite easy to prove, while the fourth is available to us if we apply what we know of set theory diligently and delicately.

Assertion 1—For $\omega_i \subset \Sigma$, $\psi(\omega_i) \geq 0$.

This is quite straightforward. Since any $\omega_i \subset \Sigma$ by definition is an executed analysis, it must be based on a positive number of participants and a positive number of variables. Thus, n_i and v_i are ≥ 0. Therefore, $\psi(\omega_i) = n_i v_i \geq 0$.

Assertion 2—$\psi(\varnothing) = 0$.

If $\omega_i = \varnothing$, the ith analysis has not been conducted. Thus, the content of the analysis is vacant. Therefore, each of n_i and v_i are equal to 0 and $\psi(\omega_i) = n_i v_i = 0$.

Assertion 3—If $\omega_i \subset \omega_j$, then $\psi(\omega_i) \leq \psi(\omega_j)$.

This assertion requires the development of the idea of one analysis *containing* another analysis ($\omega_i \subset \omega_j$), a relatively unexplored concept in clinical trial methodology.

Let's begin with the notion that an analysis ω_i is contained within analysis ω_j when (1) the observations in analysis i are the same or is a subset of those used in analysis j, i.e., $\mathbf{n}_i \subseteq \mathbf{n}_j$, and (2) the variables used in analysis i are the same or is a subset of the variables used in analysis j, i.e., $\mathbf{v}_i \subseteq \mathbf{v}_j$. With this definition, $\omega_i \subset \omega_j$ then implies that $n_i \leq n_j$ and $v_i \leq v_j$, then $n_i v_i \leq n_j v_j$ and $\psi(\omega_i) \leq \psi(\omega_j)$.

However, this practical definition has consequences. One important ramification is that even though $\omega_i \subset \omega_j$, the two analyses can be completely unrelated to each other.

For example, an analysis conducted to assess whether the change in blood pressure over time is impacted by a change over time in quality of life can be assessed using a general linear model.

However, the analysis of whether the change in quality of life over time is related to the change in blood pressure over time is also addressed by a general linear model with the same participants and variables. Both analyses utilize the same number of observations and variables and therefore each contains the other; however, the questions motivating the analyses are different. This is another example of how analysis intent is separate and apart from analysis content and must therefore be considered independently (as we will).

In addition, $\psi(\omega_i) \leq \psi(\omega_j)$ does not imply that $\omega_i \subset \omega_j$. Consider the situation where analysis ω_1 consists of 100 observations and 10 variables. Its content will be greater than that of an analysis ω_2 from the same trial that utilizes 75 observations and 5 variables. However, if the 10 variables utilized in analysis ω_1 are only demographic variables while analysis ω_2 utilizes imaging variables only, the analyses do not contain one another even though $\psi(\omega_1) \leq \psi(\omega_2)$.

Assertion 4—Let $\omega_i, i = 1,,,n$ be a collection of analyses contained in Σ. Then $\psi\left(\bigcup_{i=1}^{\infty} \omega_i\right) \leq \sum_{i=1}^{\infty} \psi(\omega_j)$.

This assertion will be demonstrated for two mutually exclusive and exhaustive cases.

Case 1. $\{\omega_1, \omega_2, \omega_3, ..., \omega_n, ...\}$ are pair-wise disjoint.
This demonstration is quite straightforward. The disjoint assumption permits $\psi(\omega_1 \cup \omega_2)$ to be written as $n_1 v_1 + n_2 v_2 = \psi(\omega_1) + \psi(\omega_2)$ since $\psi(\omega_1 \cap \omega_2) = 0$. In order to demonstrate assertion 4 in this circumstance, we simply need to add additional disjoint analyses one at a time. Adding each additional analysis ω_i into the union only adds the terms $n_i v_i$ to the content.

This can be demonstrated through induction. Assume that $\psi\left(\bigcup_{i=1}^{k}\omega_i\right)=\sum_{i=1}^{k}\psi(\omega_i).$, then develop $\psi\left(\bigcup_{i=1}^{k+1}\omega_i\right)$ as

$$\psi\left(\bigcup_{i=1}^{k+1}\omega_i\right)=\psi\left(\left(\bigcup_{i=1}^{k}\omega_i\right)\cup\omega_{k+1}\right)=\psi\left(\bigcup_{i=1}^{k}\omega_i\right)+\psi(\omega_{k+1})-\psi\left(\left(\bigcup_{i=1}^{k}\omega_i\right)\cap\omega_{k+1}\right)$$

$$=\sum_{i=1}^{k+1}\psi(\omega_i)-\sum_{i=1}^{k}\psi(\omega_i\cap\omega_{k+1})=\sum_{i=1}^{k+1}\psi(\omega_i)-0=\sum_{i=1}^{k+1}\psi(\omega_i)=\sum_{i=1}^{k+1}n_iv_i.$$

and the assertion is demonstrated for pair-wise disjoint analyses.

Case 2. $\{\omega_1,\omega_2,\omega_3,...,\omega_n,....\}$ are not pair-wise disjoint. What made case 1 so easy was the disjoint nature of the analyses under consideration. We will apply a similar approach to case 2, but we will need to essentially convert nondisjoint sets to disjoint sets so that the union of the two is the same. Then we can operate on the union of the disjoint sets. This set methodology will be a common motif in our subsequent development of quanta analysis.

Step 1: Converting nondisjoint sets to disjoint sets.
Let's build our union of sets up one at a time, beginning with $n = 2$. For this simple case, we know that $\psi(\omega_1\cup\omega_2)=\psi(\omega_1)+\psi(\omega_2)-\psi(\omega_1\cap\omega_2)$. Now suppose we looked at this differently, constructing $\omega_1\cup\omega_2$ from all of ω_1 and the part of ω_2 that does not contain ω_1,—i.e., in this formulation, $B_1=\omega_1$ and $B_2=(\omega_1\cup\omega_2)\cap\omega_1^c$., it is the $(\omega_1\cup\omega_2)\cap\omega_1^c$. This is the key (figure 1).

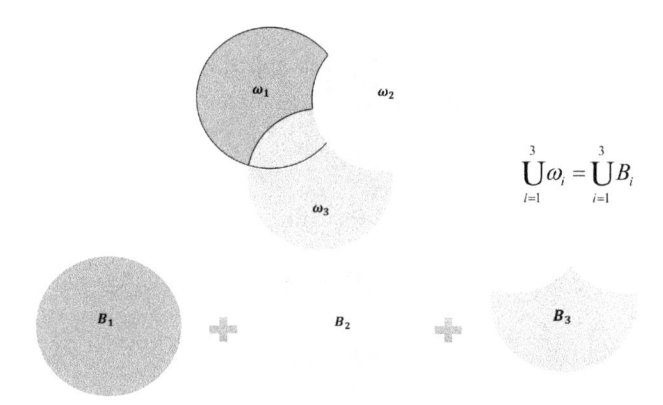

Figure 1. Decomposing the overlapping analyses $\{\omega_1, \omega_2, \omega_3\}$ into non-overlapping analysis components $\{B_1, B_2, B_3\}$

$$\bigcup_{i=1}^{3}\omega_i = \bigcup_{i=1}^{3}B_i$$

Let's first demonstrate that $B_1 \cup B_2 = \omega_1 \cup \omega_2$. This is quickly accomplished:

$$B_1 \cup B_2 = \omega_1 \cup \left(\left(\omega_1 \cup \omega_2\right) \cap \omega_1^c\right)$$
$$= \left(\omega_1 \cup \left(\omega_1 \cup \omega_2\right)\right) \cap \left(\omega_1 \cup \omega_1^c\right)$$
$$= \left(\omega_1 \cup \omega_2\right) \cap \Omega$$
$$= \left(\omega_1 \cup \omega_2\right)$$

So the unions are the same, but are B_1 and B_2 disjoint? Write

$$B_1 \cap B_2 = \omega_1 \cap \left(\left(\omega_1 \cup \omega_2\right) \cap \omega_1^c\right)$$
$$= \left(\omega_1 \cap \left(\omega_1 \cup \omega_2\right)\right) \cap \left(\omega_1 \cap \omega_1^c\right)$$
$$= \omega_1 \cap \varnothing$$
$$= \varnothing.$$

Thus, $\psi(\omega_1 \cup \omega_2) = \psi(B_1 \cup B_2) = \psi(B_1) + \psi(B_2)$, since B_1 and B_2 are disjoint. This approach will permit us to manage more complicated unions of n nondisjoint analyses, as follows.

From $\{\omega_i\}\, i = 1,...,n$ nondisjoint analyses, we want to find $\psi\left(\bigcup_{i=1}^{n} \omega_i\right)$. We begin by defining a new sequence of increasing sets $\{C_i\}$ as $C_1 = \omega_1$, $C_2 = \omega_1 \cup \omega_2$, $C_3 = \omega_1 \cup \omega_2 \cup \omega_3,...$, etc. Note that $\bigcup_{i=1}^{n} \omega_i = C_n = \bigcup_{i=1}^{n} C_i$.

Now, as in our first example, we define a collection of sets $\{B_i\}$, such that $B_i = C_i \cap C_{i-1}^c$. Thus, $B_1 = C_1 = \omega_1$, $B_2 = C_2 \cap C_1^c$, $B_3 = C_3 \cap C_2^c$, etc.

Note that the union of sets $\{B_1\}$ recapitulates $\{C_1\}$. For example,
$$B_1 \cup B_2 = C_1 \cup (C_2 \cap C_1^c) = (C_1 \cup C_2) \cap (C_1 \cup C_1^c)$$
$$= C_2 \cap \Omega = C_2,$$

and
$$(B_1 \cup B_2) \cup B_3 = C_2 \cup (C_3 \cap C_2^c)$$
$$= C_2 \cup C_3 \cap (C_2 \cup C_2^c) = C_3.$$

This logic can be extended to show that $\bigcup_{i=1}^{n} \omega_i = \bigcup_{i=1}^{n} C_i = \bigcup_{i=1}^{n} B_i$ and therefore
$$\psi\left(\bigcup_{i=1}^{n} \omega_i\right) = \psi\left(\bigcup_{i=1}^{n} C_i\right) = \psi\left(\bigcup_{i=1}^{n} B_i\right).$$

Weighing the Evidence

However, the members of the set $\{B_i\}$ are pair-wise disjoint.[*] Therefore,

$$\psi\left(\bigcup_{i=1}^{n}\omega_i\right) = \psi\left(\bigcup_{i=1}^{n}B_i\right) = \sum_{i=1}^{n}\psi(B_i).$$

The formula $\psi\left(\bigcup_{i=1}^{n}\omega_i\right) = \sum_{i=1}^{n}\psi(B_i)$ is critical for us. It formulates the content of the union of a collection of analyses into the simple sum of the content of disjoint combinations of analyses. The set $\{B_i\}$ represent a collection of analysis fragments or quanta that represent separate disjoint contributions to the union of all n analyses.

Demonstration of assertion 4.

Attention turns to $\sum_{i=1}^{n}\psi(B_i)$ in order to demonstrate the veracity of assertion 4, i.e., that $\psi\left(\bigcup_{i=1}^{n}\omega_i\right) \leq \sum_{i=1}^{n}\psi(\omega_j)$.

Let's begin with the simplest case first. For $n = 1$, since $B_1 = \omega_1$, write $\psi(B_1) = \psi(\omega_1)$.

---

[*] This follows from $B_j \cap B_i = (C_j \cap C_{j-1}^c) \cap (C_i \cap C_{i-1}^c)$ and from there

$$C_{j-1}^c = \left(\bigcup_{m=1}^{i-1}C_m \cup C_i \cup \bigcup_{m=i+1}^{j-1}C_m\right)^c = \underset{m=1}{\overset{i-1}{\mathrm{I}}}C_m^c \cap C_i^c \cap \underset{m=i+1}{\overset{j-1}{\mathrm{I}}}C_m^c.$$

Thus,

$$B_j \cap B_i = C_j \cap C_{j-1}^c \cap C_i \cap C_{i-1}^c = C_j \mathrm{I}\left(\underset{m=1}{\overset{i-1}{\mathrm{I}}}C_m^c \cap C_i^c \cap \underset{m=i+1}{\overset{j-1}{\mathrm{I}}}C_m^c\right) \cap C_i \cap C_{i-1}^c$$

$$= C_j \mathrm{I}\underset{m=1}{\overset{i-1}{\mathrm{I}}}C_m^c \cap \underset{m=i+1}{\overset{j-1}{\mathrm{I}}}C_m^c \cap C_{i-1}^c \cap C_i^c \cap C_i = \emptyset.$$

For $n = 2$, write

$$\psi\left(\bigcup_{i=1}^{2}\omega_i\right) = \sum_{i=1}^{2}\psi(B_i) = \psi(B_1) + \psi(B_2)$$
$$= \psi(\omega_1) + \psi(B_2).$$

Recall that
$$B_2 = C_2 \cap C_1^c = (\omega_1 \cup \omega_2) \cap \omega_1^c$$
$$= (\omega_1 \cap \omega_1^c) \cup (\omega_2 \cap \omega_1^c) = \omega_2 \cap \omega_1^c.$$

However, $\omega_2 \cap \omega_1^c \subseteq \omega_2$, implying that $\psi(\omega_2 \cap \omega_1^c) \leq \psi(\omega_2)$ by assertion 3. Now also note that

$$\psi(B_2) = \psi(C_2) - \psi(C_1)$$
$$= \psi(\omega_1 \cup \omega_2) - \psi(\omega_1)$$
$$= n_1 v_1 + n_2 v_2 - n_{12} v_{12} - n_1 v_1$$
$$= n_2 v_2 - n_{12} v_{12}.$$

Substituting the actual measure values, we have $\psi(B_2) = n_2 v_2 - n_{12} v_{12}$.

If there are no observations in common, then $n_{12} = 0$, and $\psi(\omega_1 \cup \omega_2) = \psi(B_1) + \psi(B_2) = n_1 v_1 + n_2 v_2$ and we have the equality. If the analyses are not disjoint, then
$$\psi(\omega_1 \cup \omega_2) = \psi(B_1) + \psi(B_2) = n_1 v_1 + n_2 v_2 - n_{12} v_{12}$$
$$< \psi(\omega_1) + \psi(\omega_2).$$

Thus
$$\psi\left(\bigcup_{i=1}^{2}\omega_i\right) = \psi(B_1) + \psi(B_2) \leq \psi(\omega_1) + \psi(\omega_2).$$

This same logic applies for $n = 3$. Since it has been demonstrated that $\psi\left(\bigcup_{i=1}^{3}\omega_i\right) = \sum_{i=1}^{3} B_3$ and also that

$$\psi\left(\bigcup_{i=1}^{2}\omega_i\right) = \psi(B_1) + \psi(B_2) \leq \psi(\omega_1) + \psi(\omega_2),$$

then it must only be shown that $\psi(B_3) \leq \psi(\omega_3)$. This follows from

$$B_3 = C_3 \cap C_2^c = (\omega_1 \cup \omega_2 \cup \omega_3) \cap (\omega_1 \cup \omega_2)^c$$

$$= \left((\omega_1 \cup \omega_2) \cap (\omega_1 \cup \omega_2)^c\right) \cup \left(\omega_3 \cap (\omega_1 \cup \omega_2)^c\right)$$

$$= \omega_3 \cap (\omega_1 \cup \omega_2)^c \subseteq \omega_3.$$

Implementing ψ-content, write

$$\psi(B_3) = \psi(C_3) - \psi(C_2) = \psi(\omega_1 \cup \omega_2 \cup \omega_3) - \psi(\omega_1 \cup \omega_2)$$

$$= \psi(\omega_1 \cup \omega_2) + \psi(\omega_3) - \psi\left((\omega_1 \cup \omega_2) \cap \omega_3\right) - \psi(\omega_1 \cup \omega_2)$$

$$= \psi(\omega_3) - \psi\left((\omega_1 \cup \omega_2) \cap \omega_3\right).$$

Continuing, since

$$\psi\left((\omega_1 \cup \omega_2) \cap \omega_3\right) = \psi\left((\omega_1 \cap \omega_3) \cup (\omega_2 \cap \omega_3)\right)$$

$$= \psi(\omega_1 \cap \omega_3) + \psi(\omega_2 \cap \omega_3) - \psi(\omega_1 \cap \omega_2 \cap \omega_3)$$

$$= n_{13}v_{13} + n_{23}v_{23} - n_{123}v_{123}$$

then

$$\psi(B_3) = \psi(\omega_3) - \psi\left((\omega_1 \cup \omega_2) \cap \omega_3\right)$$

$$= n_3 v_3 - n_{13}v_{13} - n_{23}v_{23} + n_{123}v_{123}.$$

This can be written as

$$\psi(B_3) = n_3 v_3 - (n_{13}v_{13} - n_{123}v_{123}) - n_{23}v_{23}$$

since $n_{123}v_{123} \leq n_{13}v_{13}$ and $n_{23}v_{23} \geq 0$, Thus, $\psi(B_3) \leq \psi(\omega_3) = n_3 v_3$.

and $\psi\left(\bigcup_{i=1}^{3}\omega_1\right) \le \sum_{i=1}^{3}\psi(\omega_3)$.

An induction argument is available. Assume $\sum_{i=1}^{k}\psi(B_i) \le \sum_{i=1}^{k}\psi(\omega_i)$. To show that $\sum_{i=1}^{k+1}\psi(B_i) \le \sum_{i=1}^{k+1}\psi(\omega_i)$, we only need show that $\psi(B_{k+1}) \le \psi(\omega_{k+1})$. This we do by writing

$$B_{k+1} = C_{k+1} \cap C_k^c = (C_k \cup \omega_{k+1}) \cap C_k^c$$
$$= \omega_{k+1} \cap C_k^c \subseteq \omega_{k+1}.$$

So $\psi(B_{k+1}) \le \psi(\omega_{k+1})$, $\sum_{i=1}^{k+1}\psi(B_i) \le \sum_{i=1}^{k+1}\psi(\omega_i)$, and

$$\psi\left(\bigcup_{i=1}^{k+1}w_i\right) = \sum_{i=1}^{k+1}\psi(B_i) \le \sum_{i=1}^{k+1}\psi(\omega_i).$$

Thus, assertion 4 is proven, and ψ is a countably additive measure.

Chapter Summary

We have spent considerable time developing the motivation for and exploring the limitations of ψ-content. While ψ is certainly a measurable function on the analysis space (Ω, Σ), we need more than that from it. We require the ability to compute the content of a region of analyses in a way that makes some intuitive sense.

What defines *sense* here is the four properties of measure. Thus, $\psi(\omega_i) = n_i v_i$ must not just be a measurable function, but a measure. Having satisfied the four properties of measure, we are assured that this is the case and can proceed with computing the measure of analysis regions with confidence since the properties of a measure are the characteristics that we need to gain an intuitive sense of analysis regions.

Measuring Analysis Sets (Quanta Analysis)

The ultimate goal of this overall development is to integrate over a collection of analysis regions A_q a measurable function $f(\omega_i)$ with respect to ψ-measure, $\int_{A_q} f(\omega_i) d\psi$, a procedure that will manage the redundancy in the observations and variables in the different analyses that comprise A_q. If, for example, the investigators wish to assess the totality of evidence in a clinical trial that addressed question q ("What is the beneficial effect of therapy in Asian women?"), they would deploy an especially derived measurable function $f(\omega_i)$ that assessed benefit and then compute $\int_{A_q} f(\omega_i) d\psi$, where $A_q = \{\omega_i / q_i @ q\}$. However, this integral is difficult to compute directly because the analyses are in general not disjoint, and therefore the veracity of $\int_A f(\omega_i) d\psi = \sum_{\omega_i \subset A} f(\omega_i) \psi(\omega_i)$ cannot be assumed.* However, from the statement that $\bigcup_i \omega_i = A_q$ (a statement about the analysis region) and (1) our previous construction of the collection of increasing sets $\{C_i\}$, such that $C_i = \bigcup_{j=1}^{i} \omega_1$, and (2) the collection of disjoint sets $\{B_i\}$, such that $B_i = C_i \cup C_{i-1}^c$, and $\bigcup_i B_i = A_q$, then the quantity $\int_{A_q} f(\omega_i) d\psi$ can be computed exactly as

---

* In fact, we know that $\int_{A_q} d\psi = \psi\left(\bigcup_{\omega_i \subset A_q} \omega_i\right) \leq \sum_{\omega_i \subset A_q} \psi(\omega_i)$ by the fourth property of measure theory discussed earlier.

$$\int_{A_q} f(\omega_i)d\psi = \int_{\bigcup_j B_i} f(\omega_i)d\psi = \sum_{B_i \subset A} f(\omega_i)\psi(B_i).$$ The construction of the disjoint sets $\{B_i\}$ is what converts the integral into a summation.

The sets $\{B_i\}$ are what we call the fractions or quanta of analyses. It only remains to compute the measure of each quanta, $\psi(B_i)$, the topic of this chapter.

Computing Quanta Sums

Beginning with B_1, write $\psi(B_1) = \psi(\omega_1) = n_1 v_1$ as we saw earlier. For $i = 2$, we write
$$\psi(B_2) = \psi(C_2) - \psi(C_1)$$
$$= \psi(\omega_i \cup \omega_2) - \psi(\omega_1)$$
$$= n_1 v_1 + n_2 v_2 - n_{12} v_{12} - n_1 v_1$$
$$= n_2 v_2 - n_{12} v_{12}.$$

Recall that this finding was a basis of the observation that $\psi(B_2) \leq \psi(\omega_2)$, and $\psi(\omega_1 \cup \omega_2) \leq \psi(\omega_1) + \psi(\omega_2)$.

For $i = 3$,
$$\psi(B_3) = \psi(C_3) - \psi(C_2) = \psi(\omega_1 \cup \omega_2 \cup \omega_3) - \psi(\omega_1 \cup \omega_2)$$
$$= \psi(\omega_1 \cup \omega_2) + \psi(\omega_3) - \psi((\omega_1 \cup \omega_2) \cap \omega_3) - \psi(\omega_1 \cup \omega_2)$$
$$= \psi(\omega_3) - \psi((\omega_1 \cup \omega_2) \cap \omega_3).$$

Continuing, since
$$\psi((\omega_1 \cup \omega_2) \cap \omega_3) = \psi((\omega_1 \cap \omega_3) \cup (\omega_2 \cap \omega_3))$$
$$= \psi(\omega_1 \cap \omega_3) + \psi(\omega_2 \cap \omega_3) - \psi(\omega_1 \cap \omega_2 \cap \omega_3)$$
$$= n_{13} v_{13} + n_{23} v_{23} - n_{123} v_{123}$$

then
$$\psi(B_3) = \psi(\omega_3) - \psi\left((\omega_1 \cup \omega_2) \cap \omega_3\right)$$
$$= n_3 v_3 - n_{13} v_{13} - n_{23} v_{23} + n_{123} v_{123}.$$

We see a pattern beginning to emerge. For B_3, we start with $\psi(\omega_3)$, then subtract off the measure of the dual interactions that involve ω_3, then add the triple interaction.

Continuing for $\psi(B_4)$ (the measure of the union of the analyses ω_1, ω_2, ω_3, ω_4 after removing the union of analyses ω_1, ω_2, ω_3, proceed as

$$\psi(B_4) = \psi(C_4) - \psi(C_3) = \psi(\omega_1 \cup \omega_2 \cup \omega_3 \cup \omega_4) - \psi(\omega_1 \cup \omega_2 \cup \omega_3)$$
$$= \psi(\omega_1 \cup \omega_2 \cup \omega_3) + \psi(\omega_4) - \psi\left((\omega_1 \cup \omega_2 \cup \omega_3) \cap \omega_4\right)$$
$$- \psi(\omega_1 \cup \omega_2 \cup \omega_3)$$
$$= \psi(\omega_4) - \psi\left((\omega_1 \cup \omega_2 \cup \omega_3) \cap \omega_4\right)$$
$$= \psi(\omega_4) - \psi(\omega_1 \cap \omega_4) - \psi(\omega_2 \cap \omega_4) - \psi(\omega_3 \cap \omega_4)$$
$$+ \psi(\omega_1 \cap \omega_2 \cap \omega_4) + \psi(\omega_1 \cap \omega_3 \cap \omega_4) + \psi(\omega_2 \cap \omega_3 \cap \omega_4)$$
$$- \psi(\omega_1 \cap \omega_2 \cap \omega_3 \cap \omega_4).$$

Making a simplification of notation of $(nv)_{ij...}$ for $n_{ij...} v_{ij...}$, write

$$\psi(B_4) = \psi(n_4 v_4) - \psi\left((nv)_{14}\right) - \psi\left((nv)_{24}\right) - \psi\left((nv)_{34}\right)$$
$$+ \psi\left((nv)_{124}\right) + \psi\left((nv)_{134}\right) + \psi\left((nv)_{234}\right) - \psi\left((nv)_{1234}\right)$$
$$= \psi\left((nv)_4\right) - \sum_{j=1}^{3} \psi\left((nv)_{j4}\right) + \sum_{j_2=2}^{3} \sum_{j_1=1}^{j_2-1} \psi\left((nv)_{j_1 j_2 4}\right) - \psi\left((nv)_{1234}\right).$$

Note that $(nv)_{14} \geq (nv)_{124}$, $(nv)_{34} \geq (nv)_{134}$, and $(nv)_{24} \geq (nv)_{234}$, thus $\psi(B_4) \leq n_4 v_4 = \psi(\omega_4)$.

There is an induction argument here. In general,

$$\psi(B_k) = \psi(n_k v_k) - \sum_{j_1=1}^{k-1} \psi\left((nv)_{j_1 k}\right) + \sum_{j_1=1}^{k-1} \sum_{j_2=2}^{j_2-1} \psi\left((nv)_{j_1 j_2 k}\right)$$
$$- \sum_{j_1=1}^{j_2-1} \sum_{j_2=2}^{j_3-1} \sum_{j_3=3}^{k-1} \psi\left((nv)_{j_1 j_2 j_3 k}\right) + ...$$

Thus, the measure of the collection of nondisjoint analyses $\psi\left(\bigcup_{\omega_i \subset A} \omega_i\right)$ can be assembled into the sum of mutually disjoint combinations of the measures of analysis quanta, which themselves are the measures of discrete fragments of analyses components, permitting $\int_{A_q} d\psi = \sum_{\omega_i \subset A_q} \psi(B_i)$ where $\sum_{\omega_i \subset A_q} B_i = \bigcup_{\omega_i \subset A_q} \omega_i$ and thus

$$\int_{A_q} f(\omega_i) d\psi = \sum_{B_i \subset A_q} f(\omega_i) \psi(B_i).$$

Strategy in Calculating Quanta

The fact that the quantity $\psi(B_k)$ is itself composed of alternating sums and differences of the intersections of increasing numbers of measures of analysis quanta helps us in its calculation. Specifically, $\psi(B_k)$ is the measure of ω_k, minus the sum of all the dual analysis interactions that involve ω_k, plus the sum of the measure of each of the triple interactions that involve ω_k, minus the sum of the fourth level interactions that involve ω_k, and so on.[*]

As an example, in order to compute $\psi(B_7)$, first subtract from $\psi(\omega_7)$ the 6 binary interactions that involve analysis ω_7, then add the 15 triple intersections terms that include ω_7, then subtract the 20 fourth-level interactions that involve ω_7, etc. This process of term counting provides a simple way to compute $\psi(B_k)$ when $\psi\left(\bigcap_{i=1}^{n} \omega_i\right)$ is same constant for all n above some value, as demonstrated in examples in some of the later chapters. In addition, the computation is eased when the magnitude of the higher order interactions decreases, e.g., when $\lim_{n \to \infty} \psi\left(\bigcap_{i=1}^{n} \omega_i\right) = \lim_{n \to \infty}(n_{123...n} v_{123...n}) = 0.$[†]

---

[*] The number of terms that comprise each of the levels of interactions for $\psi(B_k)$ are generated from the kth row of Pascal's triangle.

[†] Since there is no clinical research effort with an infinite number of participants and variables, this limit argument is not as helpful as we might like. The

Weighing the Evidence

It's now time for a welcome review of where we are.

---

smallest nonzero value $\psi\left(\prod_{i=1}^{n} \omega_i\right) = n_{123...n} v_{123...n}$ can be is one since an analysis, in order to have positive measure, must have at least one observation and one variable.

A Breather …

The last chapter covered a lot of new material, so let's pause for a moment and recapitulate.

We are interested in measuring the content of analysis. The purpose of this measurement is to provide an assessment of the accumulation of content over a region of analyses in a clinical research effort that addresses a clinical question q.

Since we want to take advantage of many features measure has to offer (including the ability to assess the content of overlapping analyses), we have selected a content function (working to keep it tractable for computing). We have defined the analysis content as the product of the number of participants and the number of variables used in the analysis.

Later we found that this function of an analysis, ψ, was not just a content but also a formal measure. This permits us to use ψ as a set function, with some very useful properties (e.g., the ability to compute it on unions of analyses).

Using our set theory, we saw that when the collection of analyses $\{\omega_1, \omega_2, \omega_3 \ldots \omega_n\}$ are disjoint, we could write $\psi\left(\bigcup_{i=1}^{n} \omega_i\right) = \sum_{i=1}^{n} n_i v_i$. However, our recognition that collections of analyses most frequently have shared participant and variable data, in combination with the work of the previous chapter, reveals that when the collections of analyses are nondisjoint, then $\psi\left(\bigcup_{i=1}^{n} \omega_i\right)$ can be assembled into the sum of mutually disjoint ψ-measures, not of analyses but of analyses fragments or quanta

$\{B_1, B_2, B_3 \ldots B_n\}$. Since these analysis quanta are disjoint, we can write $\psi\left(\bigcup_{i=1}^{n} \omega_i\right) = \sum_{i=1}^{n} \psi(B_i)$, where

$$\psi(B_i) = \psi(n_i v_i) - \sum_{j_1=1}^{i-1} \psi\left((nv)_{j_1 i}\right) + \sum_{j_1=1}^{j_2-1} \sum_{j_2=2}^{i-1} \psi\left((nv)_{j_1 j_2 i}\right)$$
$$- \sum_{j_1=1}^{j_2-1} \sum_{j_2=2}^{j_3-1} \sum_{j_3=3}^{i-1} \psi\left((nv)_{j_1 j_2 j_3 i}\right) + \ldots$$

We will come back to this computation in a moment. However, focusing on the big picture, we can now answer the question of what is the ψ-measure of a collection of analyses that addresses a clinical research question q. This operation is simply the accumulation of measure.

Recall that such an accumulation is signaled through the use of the integral sign. Thus, the integral $\int_{A_q} d\psi$ is the statement of our intent to accumulate the content of the set of analyses A_q using ψ-measure. Thus, $\psi(A_q) = \int_{A_q} d\psi$. However, now that we have divided the analyses into disjoint analysis fractions or quanta, we can go one step further and write

$$\psi(A_q) = \int_{A_q} d\psi = \sum_{\omega_i \subset A} \psi(B_i).$$

The content measure of the analysis set A_q is simply the sum of the ψ-measures of the analyses quanta into which the set $A_q = \{\omega_i\}$ is fractionated. If the analyses are pair-wise disjoint, then the equation simplifies to $\psi(A_q) = \sum_{\omega_i \subset A} n_i v_i$.

Some Helpful Observations

Let's make some observation before we proceed.

First, since analyses in clinical research effort commonly use the same subjects of observations and variables, it will be the rare collection of analyses that are pair-wise disjoint. Thus, the equation on which we will rely on will be $\psi(A_q) = \int_{\omega_i \subset A_q} d\psi = \sum_{\omega_i \subset A_q} \psi(B_i)$.

In addition, the computation of $\psi(B_i)$ may appear complicated, but it is simply adding and subtracting sums of participant-variable products in a specific sequence. This is readily done in a program like Excel, so we will not let this calculation impede our examination of this content measure's performance.

With this development behind us, we are now in a position to compute not just the total content of a collection of analyses $\{\omega_1, \omega_2, \omega_3, ... \omega_n\}$ as $\psi\left(\bigcup_{A_q} \omega_i\right) = \psi\left(\bigcup_{i=1}^{n} \omega_i\right) = \sum_{i=1}^{n} \psi(B_i)$, but we can also compute the fraction of the total content of the set of analyses contained by the ith analysis, $\dfrac{\psi(B_i)}{\sum_{i=1}^{n} \psi(B_i)}$. This proportion can function as a weight to be used when we want to assess the role of function of the analysis.

This is useful because ultimately what we are interested is not just in $\psi(A) = \int_{A_q} d\psi = \sum_{\omega_i \subset A} \psi(B_i)$, but in $\int_{A_q} f(\omega_i) d\psi = \sum_{\omega_i \subset A} f(\omega_i) \psi(B_i)$, where the function $f(\omega_i)$ reflects, for example, the duality of an analysis result. In this case, the equation of interest will be

$$\dfrac{\int_{\omega_i \subset A} f(\omega_i) d\psi}{\int_{\omega_i \subset A} d\psi} = \dfrac{\sum_{\omega_i \subset A} f(\omega_i) \psi(B_i)}{\sum_{\omega_i \subset A} \psi(B_i)} = \sum_{\omega_i \subset A} f(\omega_i) \left[\dfrac{\psi(B_i)}{\sum_{\omega_i \subset A} \psi(B_i)}\right]$$

In this case, the component $\dfrac{\psi(B_i)}{\sum_{\omega_i \subset A} \psi(B_i)}$ is a weighting function, either increasing or decreasing the contribution $f(\omega_i)$ based on the proportion of the total content measure in the set of analyses A that is explained by individual analysis ω_i.

Now let's look at some simple examples.

A First Demonstration

We have, so far, developed an analysis platform specifically tailored to health care research. It is based on the ultimate source of useful information in clinical research—the participants and their variables.

We have used set theory to assure ourselves that there are clear mathematical rules governing how we manage collections of data in this participant-variable dimension. And we know enough about measure theory to understand that this ψ-content function is a measure. Thus, at least theoretically, we can compute the ψ-content of a single analysis or a collection of analyses in a reliable and consistent fashion and now refer to this analysis content as ψ-measure.

This ψ-measure permits us to compute the contribution of an analysis in our construct; this contribution will be the proportion of content measure contained in the analysis or $\dfrac{\psi(B_i)}{\int_A d\psi} = \dfrac{\psi(B_i)}{\sum_{\omega_i \subset A} \psi(B_i)}$, where the collection of sets $\{B_i\}$ represents the disjoint contributions of each analysis's content measure to the overall content of the analysis set $\int_A d\psi$, expressed as a simple proportion.

An examination of the characteristics of this measure will help us to assess its ability to identify potential contributions of different analyses to clinical research results. Our construction of the set $\{B_i\}$ in the last two chapters demonstrated that it is an individual analysis's quanta, B_i, that makes a contribution to the content measure of the union of the set of analyses $\bigcup_q \omega_i$ separate and apart from the other ω_i. It will therefore be of particular interest to determine when these quanta make substantial versus small contributions to the ψ-measure of this union.

Thus, comparing the content measures of analysis sets in particular research circumstances will reveal how to use ψ-measure and give us at least an initial sense of how its output compares with the commonly used standard.

Example—A Single Primary Endpoint

Let's begin with a randomized clinical trial with two treatment arms designed to assess the effect of a new therapy in patients with heart failure. Subjects are randomized into two treatment groups (treatment and placebo) and will have several outcome measures assessed. Each of these outcomes is prospectively declared and measured with high quality.

For example, the heart's ability to pump effectively will be directly assessed by estimators that assess ejection fraction, end systolic volume, end diastolic volume, or cardiac output. In addition, there are evaluations of the vitality of the individual (e.g., walking distance or quality of life questionnaires). Each of these measures is taken at baseline and then once again at some prespecified time in the future (e.g., one year) and their differences obtained and assessed.

In the traditional paradigm, a clinical trial focuses on one (or a small number) of endpoints or analyses (termed *primary*) even though several and sometimes many more endpoints and analyses were prospectively declared; the trial results are based in large part (and sometimes exclusively) on the magnitude of these primary analyses' *p*-values.

Thus, in a trial with a single primary endpoint selected from k-prospectively declared analyses, the remaining $k - 1$ analyses, while interesting and providing additional support for the primary endpoint's finding, in and of themselves do not formerly contribute to the result of the study. The term typically used for these additional, secondary endpoint findings is *supportive*.

Our goal is to examine the implications of the application of this ψ-measure in this example.

Assume that there are $\omega_1, \omega_2, \omega_3, ..., \omega_K$ individual analyses to be carried out in this clinical trial, one for each of the k prospective declared evaluations, and let ω_1 be the primary analysis.

Let's also assume for simplicity that the same n subjects are included in each analysis* and that each analysis consists of three variables (baseline evaluation, follow-up evaluation, and treatment assignment).

The variable denoting treatment assignment is the same across all analyses, but the endpoint measures (baseline and follow-up assessments) are different from analysis to analysis. Our goal is to compute the ψ-measure of the union of these analyses, $\psi\left(\bigcup_{i=1}^{K} \omega_i\right)$ representing the total content of all these endpoints taken together.

Using our development, we must construct a collection of sets $\{B_1, B_2, B_3, ..., B_K\}$ that are disjoint as in the previous section and for which $\psi\left(\bigcup_{i=1}^{K} B_i\right) = \psi\left(\bigcup_{i=1}^{K} \omega_i\right)$.

Proceeding, for the first analysis, there are n subjects and three variables (baseline evaluation, follow-up evaluation, and treatment assignment). Thus, from our formula for ψ-measure, compute $\psi(B_1) = \psi(\omega_1) = n_1 v_1 = 3n$.

In order to compute $\psi(B_2)$, begin with $\psi(B_2) = n_2 v_2 - n_{12} v_{12} = 3n - n_{12} v_{12}$; , n_{12} being the number of observations common to the first two analyses—which, in this example, is n—and $v_{12} = 1$ since there is only one variable that is common to both analyses (the treatment group assignment). Therefore, we can compute $\psi(B_2) = n_2 v_2 - n_{12} v_{12} = 3n - n = 2n$.

Let's pause for a moment. Recall that $\psi(B_2)$ is the contribution of analysis ω_2 to the content measure of $\omega_1 \cup \omega_2$, $\psi(\omega_1 \cup \omega_2)$, separate and apart from the contribution of analysis ω_1. Note that $\psi(\omega_2) = 3n > \psi(B_2)$—that is, the measure of analysis 2 is greater than its measure when its content commonality with analysis 1 is removed.

Continuing, write
$$\psi(B_3) = \psi(\omega_3) - \psi(\omega_1 \cup \omega_3) - \psi(\omega_2 \cup \omega_3) + \psi(\omega_1 \cup \omega_2 \cup \omega_3)$$
$$= n_3 v_3 - n_{13} v_{13} - n_{23} v_{23} + n_{123} v_{123}$$
$$= n(3-1-1+1) = 2n.$$

---

* The reality of subjects with missing values is easily incorporated in ψ-measure.

In fact, $\psi(B_j) = 2n$ for all $j > 1$.* Thus, each subsequent analysis after the first contributes a separate ψ-measure or quantum of $2n$.

We are now ready to compute that

$$\psi\left(\bigcup_{i=1}^{k} \omega_j\right) = \sum_{i=1}^{k} \psi(B_j)$$

$$= 3n + \sum_{k=2}^{n} 2n = 3n + 2(k-1)n = (2k+1)n.$$

Typically, in clinical trial analyses, the only endpoint that contributes to a quantitative estimate of the effect of therapy is the primary endpoint. However, in this formulation, positive measure is available to serve as the basis for a mathematical contribution for each of the k–prospectively declared endpoints regardless which one is primary.

In fact, for three prospectively declared endpoints, the total content measure is $3n + 2n + 2n = 7n$; the fraction of the total content measure contained by the primary endpoint is $3n/7n = 0.429$, with 28.6% content measure remaining in each of the two remaining endpoints.

Thus, if this were a trial with 87 subjects, a primary outcome of the placebo-corrected effect of left ventricular end systolic volume and two prospectively declared secondary outcomes of (a) the placebo-adjusted change in left ventricular end systolic volume and (b) the placebo-adjusted change in left ventricular end diastolic volume, then $\psi(\omega_1) = (3)(87) = 261$,; and we can also compute $\psi(\omega_2) = \psi(\omega_3) = (2)(87) = 174$.

Note again that the raw content measure of an analysis ω_i, $\psi(\omega_i) = n_i v_i$ can be different from that analysis's contribution to $\psi\left(\bigcup_{i=1}^{k} \omega_j\right)$. While $\psi(B_1) = \psi(\omega_1)$ by definition, $\psi(B_2) = 2n \ne \psi(B_1) = 3n$. This is because $\psi(B_2)$ reflects the contribution of analysis ω_2 after taking ω_1 into account. Similarly,

---

* As we have seen, the quantity $\psi(B_k)$ is itself composed of alternating sums and differences of the intersections of increasing numbers of measures of analysis quanta. This process of term counting provides a simple way to compute $\psi(B_k)$, when $\psi\left(\bigcap_{i=1}^{n} \omega_i\right)$ is same constant for all n above some value, as demonstrated in the examples.

$\psi(B_i)$ is the measure of ω_i after taking into account the sequence of analyses $\omega_1, \omega_2, \omega_3, ...\omega_{i-1}$.

Initial Analysis Sequencing Observation

In the previous example with three outcomes and no missing data, we find that, however, the three outcomes are sequenced $\psi\left(\bigcup_{i=1}^{3}\omega_j\right) = 7n = \sum_{i=1}^{3}\psi(B_i)$. Each of these two computations is sequence invariant.

However, the value of $\psi(B_i)$ is not. It is easy to see from this simple example that the contribution of the quantum $\psi(B_i)$ is either $3n$ or $2n$ depending on where in the sequence of analyses, $\omega_1, \omega_2, \omega_3, ..., \omega_n$ the analysis ω_i resides.

This finding has important implications for the use of quanta analysis and will now be examined in detail.

Analysis Priorities and Quanta Paths

Our first demonstration of quanta analysis applied to a simple, clinical trial scenario provided some intriguing findings.

First, not all content measure is subsumed by the primary outcome. In fact, in the previous example with 87 subjects and 3 prospectively declared outcomes, more than 50% of the total content measure of all three analyses resides with the two nonprimary evaluations.

This observation opens the door to the possibility that prospectively declared, nonprimary analyses may also be quantitatively considered in summarizing a trial result.

The current state-of-the-art analysis paradigm operates as though all "analysis measure" is absorbed by the primary evaluation. In this traditional framework, the secondary outcomes may, of course, be cerebrally integrated into the final result, but they are treated as though they "have measure zero."

The quanta analysis opens the door to the possibility of a quantitative combination of primary and secondary outcomes, a concept that will be carefully cultivated and developed later in this book.

Sequencing Variant Quanta Values

A second observation is that the contribution of an individual analysis to the content measure contained by the union of all analyses depends on where in the sequence of evaluations the particular analysis lies. Thus, while the overall measure of the union of all analyses is the same regardless of sequencing, the contribution that any particular analysis makes to that cumulative content measure depends on where in the sequence of analyses it is located.

This is an important new finding of the quanta approach, which, at first blush, takes us aback.

Having an analysis's quantum contribution be based on the sequence of analyses means that there is no unique "solution" to the value of $\psi(B_i)$. If, for example, there are three prospectively declared outcomes—the difference in the change in each of (1) left ventricular ejection fraction (LVEF), (2) end systolic volume (ESV), and (3) end diastolic volume (EDV)—then there are 3! = 6 possible sequences of analyses. They are

> LVEF ESV EDV
> LVEF EDV ESV
> ESV LVEF EDV
> ESV EDV LVEF
> EDV LVEF ESV
> EDV ESV LVEF

Each represents a sequence of evaluations and evaluates an outcome in a different sequence position.

How do we manage this?

This sequence-dependent value of the quantum $\psi(B_i)$ means that additional input is required to actually set it. The non-uniqueness of $\psi(B_i)$ is precisely what we need.

That input comes from the clinical investigators.

Assigning Location to Sequence Variant Analyses

It is the investigators and epidemiologists who provide the missing sequencing information.

For example, if the investigators believe that there is a clear primary endpoint that can be prospectively declared, measured precisely, and is both accepted and expected by the research and regulatory community, then this should be the first outcome in the sequence ω_1.

For all other outcomes that use the same participants and variables, then declaring the primary analysis as ω_1 gives the primary outcome the maximum measure. In our previous example, the primary endpoint of left ventricular ejection fraction would be first in the sequence, and as we computed, $\psi(B_1) = \psi(\omega_1) = 261$. The secondary endpoints each had a content measure of $\psi(B_2) = \psi(B_3) = 174$.

If not left ventricular ejection fraction but left ventricular end systolic volume was the primary endpoint, then it would be left ventricular end systolic volume that had the quanta value of 261. In this elementary example, it does not matter which is the second or third analysis outcome in the sequence because they each have the same quanta value[*].

In fact, the sequence-dependent nature of the quanta construction is a mathematical justification for the selection of the primary endpoint.

The motivations for the selection of a primary outcome is multidimensional. The epidemiologic goal in the selection of an outcome as primary is to have an outcome measure that is responsive to the intervention, can be measured with sufficient precision, and is acceptable to the research and regulatory community.

From the measure-theoretic perspective, making this selection is the same as choosing the first analysis ω_1 in the sequence of analyses; maximum content measure is assigned to it in accordance with our set theoretic rules for manipulating measure, avoiding any reduction due to intersections with other analysis sets.[†]

The primary analysis is given the optimal content measure, ceteris paribus. However, this does not mean that the primary outcome has the greatest measure. Consider the circumstance where the primary outcome consists of an analysis based on 100 participants and 3 variables, while the single secondary analysis consists of 100 participants and 8 variables, 3 of which are in common with the first analysis. Then we compute the measure for the primary outcome as $\psi(B_1) = n_i v_i = 300$, which is less than the ψ-measure for the secondary outcome, which is $\psi(B_2) = n_2 v_2 - n_{12} v_{12} = 800 - 300 = 500$. In this circumstance, the quantum for the secondary analysis exceeds that of the primary analysis.

By optimal content measure, we simply mean content measure that is not adjusted for intersections with other analysis sets, i.e.,

---

[*] These are sequent invariant quanta in this example.

[†] One might argue that the measure-theoretic approach of assigning content measure is not unlike assigning alpha prospectively. However, content measure can be assigned to a rich collection of complex intersections and unions of analyses. While probability is a measure (demonstrated by the great Russian probabilist Andrey Kolmogorov), the *p*-value (which is a specific probability of a particular event) is not, a finding that limits its flexibility as we will see in future chapters.

$\psi(B_1) = \psi(\omega_1)$.* The observations and variables that contribute to the primary analysis outcome are first made to the primary analysis. This gives the primary analysis the greatest opportunity to have a large influence on $\psi\left(\bigcup_{i=1}^{k} \omega_j\right)$. Other analyses can be considered, but only if they have positive content measure after consideration of the primary outcome.

Example—Multiple Primary Outcomes

Let's modify the first example, now affirming there are three prospectively declared competitors to be the primary outcomes: (1) the difference in the change in left ventricular ejection fraction (ΔL), (2) the difference in the change in left ventricular end systolic volume (ΔS), and (3) the difference in the change in left ventricular end diastolic volume (ΔD). We assume the same number of observations and variables as the previous example.

One traditional approach to managing this problem would require the physician investigators to select just one of them as the primary outcome, relegating the other two to playing secondary, supportive roles. This places the investigators in a difficult position because there may be no scientific justification for the selection of one of these as primary over the other two.

The alternative would be to select two or three of them as primary and distribute type I error in accordance with a multiplicity criteria such as Bonferroni. This choice increases the sample size.†

From the measure-theoretic perspective, we take a different approach. We already know that for the outcome that is the primary, we would assign content measure of 300, and to each of the other two, we would also assign 200 to produce $\psi\left(\bigcup_{i=1}^{3} \omega_i\right) = 700$. We have six such sequences

---

* This also demonstrates from a content measure approach the potential value of considering secondary outcomes with the primary analysis, which will be discussed later.

† A multiplicity criteria decreases the type I error for each endpoint selected as primary, and decreased type I error ceteris paribus increases the sample size.

of primary outcomes.* For each sequence, we will plan to conduct a full quanta analysis. Then when completed, their results will be averaged.

Specifically, if the goal of the evaluations is to compute

$$\frac{\int_{\omega_i \subset A} f(\omega_i) d\psi}{\int_{\omega_i \subset A} d\psi} = \frac{\sum_{\omega_i \subset A} f(\omega_i)\psi(B_i)}{\sum_{\omega_i \subset A} \psi(B_i)} = \sum_{\omega_i \subset A} f(\omega_i) \left[\frac{\psi(B_i)}{\sum_{\omega_i \subset A} \psi(B_i)} \right],$$

then the final equality on the right will be computed for each of the six primary outcome sequences. Thus, we can write our goal as

$$n_s^{-1} \sum_{s=1}^{n_s} \sum_{\omega_i \subset A} f(\omega_i) \left[\frac{\psi(B_{i,s})}{\sum_{\omega_i \subset A} \psi(B_{i,s})} \right],$$

where s indexes the sequences, n_s is the number of sequences that have to be considered, and $B_{i,s}$ is the quantum of the ith analysis in the sth sequence.

Since there is no priority in the outcome sequences, they should be weighted the same, permitting each sequence's impact to be equally considered.

This procedure frees investigators from having to choose a single primary outcome from among several absent criteria to inform the selection process.

This approach is easily adaptable to other circumstances. Let's modify the previous example somewhat more. Now the investigators choose a single primary outcome, and the remaining two are secondary. In the traditional paradigm, there is no hierarchy among secondary outcomes. There, the secondary outcomes do not quantitatively commit to the result of the clinical trial, so no hierarchy is required. Secondary outcomes are reported as merely supportive.

From the quanta perspective, as we have seen, secondary outcomes can provide quantitative support for the trial's answer to the clinical question that motivated it. In the circumstance where there is one

---

* $(1,2,3), (1,3,2), (2,1,3), (2,3,1), (3,1,2), (3,2,1)$

primary outcome and two secondary outcomes, one can sequence; however, the sequences are restricted.

Specifically, the primary outcome is the first outcome in any sequence, and the secondary outcomes (and only the secondary outcomes) are permuted. Thus, there are only two that must be considered (1, 2, 3), (1, 3, 2), and only these two sequences need to have their results averaged.

At What Level Does Averaging Take Place?

By computing these averages, the investigators are released from having to artificially select a primary endpoint from among candidates' primary endpoints, which are each—from a precision, culture, and sample-size perspective—essentially equivalent. It eliminates the need to make a "best guess" at what the primary endpoint should be.[*] However, we must be clear where the averaging takes place. We are not averaging ψ-measures for which there is little justification from our measure-theoretic background. Instead, we are averaging at the level of our integral $\int_{\omega_i \subset A} f(\omega_i) d\psi$, itself suitable-normed. How this operates will be clear in the discussion following the topside function development.

Multiple Manuscripts

As a final example in this chapter, consider that a collection of investigators conducts a randomized clinical trial to examine the relationship between a new lipid-lowering agent in patients at risk of having a second heart attack. In accordance with the traditional paradigm, they conduct analyses on a single primary endpoint (e.g., combined fatal and nonfatal myocardial infarction) on the overall cohort, conduct similar analyses on a small number of prospectively declared secondary outcomes, and then examine the effect of therapy on the primary outcome for a number of proper subgroups (including subjects with diabetes).[†] These results are published.

---

[*] How this operates will be demonstrated after the topside function discussion.

[†] There are also safety analyses. These will be discussed in a later chapter.

Weighing the Evidence

As is commonly the case, the investigators then decide to conduct a new sequence of analyses on the subgroup of diabetic patients. These analyses examine the effect of therapy on the primary and all secondary analyses on the diabetic subgroup, as well as on a collection of evaluations particularly focused on the impact of diabetes mellitus (e.g., amputations, stroke, and deterioration of vision). These additional results in this diabetic cohort are published in a second manuscript.

The question is, how is type I error managed across the two manuscripts? Is there an overall assessment of the type I error rate for the entire research endeavor?

The practitioners of statistical hypothesis testing are relatively mute on the application of the *p*-value arithmetic in this rather simple and common research paradigm. Even though the analyses for the second paper on the diabetic cohort was prespecified, there was no prima facie type I error for the analyses of the second paper.

If this were the case, then the investigators would have had to hold some portion of the type I error rate aside for the second paper in the traditional paradigm. But then how would this portion of the type I error have to be apportioned in the analyses on the diabetic subcohort? This would be an awkward alpha calculation at best, demonstrating the relative inflexibility of statistical inference to manage a common problem in the many clinical research efforts that each generate multiple manuscripts.

The quanta evaluation process is not interrupted simply because different collections of analyses are segregated into different publications. The investigators simply need to choose their sequence of analyses and then compute the quanta for the collection of sets $\{B_i,\ i=1,2,3...m\}$, which span the two (or more) papers. It is possible that some of the quanta that are deep in the sequence may be exceedingly small, but if there is no consensus on the sequence of analyses beyond a certain point in the sequence, then one simply uses the function

$$(m-k)\sum_{s=1}^{m-k}\sum_{\omega_i \subset A} f(\omega_i) \left[\frac{\psi(B_{i,s})}{\sum_{\omega_i \subset A} \psi(B_{i,s})} \right]$$

to average over the *m−k* analyses that are beyond the sequencing ability of the investigators. Recalling that this computation is for a study response to a question q, this response accepts the contribution of each analysis to the question's answer, regardless of which paper in which the result appears.

Subgroup Evaluations

Another situation in which the concept of analysis measure can make a contribution to the evaluation of clinical trials is in subgroup evaluations.

In a clinical trial, a subgroup analysis is the evaluation of a randomly assigned exposure on a subcohort based on strata membership determined by participant characteristics at baseline (e.g., gender or age).

These evaluations are historically fraught with concern because they can involve a small number of subjects. The strata-specific statistical hypothesis tests do not have adequate statistical power, and the type I multiplicity metric is difficult to address. Therefore, the standing rule for subgroup evaluations is to essentially set aside the subgroup group finding and be guided by the finding in the overall cohort.[1]

It would be interesting to examine this issue from the set and measure theoretic perspective.

As an example, consider a randomized clinical trial assessing the change from baseline to follow-up for a single outcome by therapy assignment. The trial conducts five analyses:

1) ω_m = the effect of therapy on males
2) ω_f = the effect of therapy on females
3) ω_w = the effect of therapy on whites
4) ω_n = the effect of therapy on nonwhites
5) ω_T = the effect of therapy on the overall cohort.

In computing the content measure of each of these analyses, the number of variables utilized in each analysis will be the same three (baseline measure, follow-up measure, and treatment identity). However, the number of observations varies from analysis to analysis. Since only males are considered in the first analysis ω_m, writing $\psi(\omega_m) = 3n_m$, where n_m denotes the number of males. Similarly, for females, find $\psi(\omega_f) = 3n_f$. Then in a straightforward manner, we can compute the necessary quanta $\psi(B_i), i = 1...5$. Observe that

$$\psi(B_m) = 3n_m$$
$$\psi(B_f) = 3n_f - 3n_{mf} = 3n_f.$$

Because there are no individuals who are both male and female and similarly no participants who are both white and nonwhite, the quanta of B_w is computed to be

$$\psi(B_w) = 3n_w - 3n_{mw} + 3n_{wmf}$$
$$= \left(3n_w - 3\left(n_{mw} + n_{fw}\right)\right) + 3n_{wmf} = 0 + 0 = 0.$$

The contribution of the white race strata is zero after considering the contribution of both gender strata. A similar result is identified for

$$\psi(B_n) = 3n_n - 3n_{mn} - 3n_{fn} + 3n_{mwn} + 3n_{mfn} - 3n_{nfwn} = 0.$$

Thus, the contribution of the two racial analyses to the accumulating union $\bigcup_{i=1}^{4} \omega_i$ is zero after considering the contribution of the two gender analyses.

However, reversing the sequence, the contribution of the two gender analyses to $\psi\left(\bigcup_{i=1}^{4} \omega_i\right)$ is zero after considering the two racial strata analyses.

It can also be shown that the contribution of the analysis of the total cohort is zero after considering either the two gender analyses or the two racial analyses. In addition, the contribution of both the gender and the race strata is zero after first considering the contribution of the total cohort. To continue this example, if the analyses are considered in the sequence $\omega_T, \omega_m, \omega_f, \omega_w, \omega_n$, then each of the gender strata and the race analyses make no contribution to the measure of $\bigcup_{i=1}^{5} \omega_i$.

Thus, as in the previous example, while $\psi\left(\bigcup_{i=1}^{5} \omega_i\right) = \sum_{i=1}^{5} \psi(B_i)$ is sequence invariant, the contribution of each of the ω_i (through its computation of B_i) to this measure is sequence dependent.

The subgroup example demonstrates that the extreme redundancy in analyses can drive particular analysis quanti B_i to zero. The operation in the subgroup example is a mathematical justification for the discounting of subgroup analyses after the overall cohort has been assessed.*

Notation

In order to incorporate the concept of a priority sequence structure formally into the measure-accumulation mechanism, define a function T that oversees the reordering of the sequence of analyses $\{\omega_i\}$, from essentially a random collection of analyses to the ordered set, to the investigator-chosen sequence of analyses. Here, $T(\omega_1, \omega_2, \omega_3, ..., \omega_n) = \omega_{[1]}, \omega_{[2]}, \omega_{[3]}, ..., \omega_{[n]}$, where the subscript $[i]$ denotes the ith analysis in the order or priority determined by the investigators. Note that the function T also converts the sequence $\{B_i\}, i = 1, 2, 3, ...$ to $\{B_{[i]}\}, i = 1, 2, 3, ...$, with this latter sequence of disjoint sets corresponding to the sequence of ordered analyses. Note that the function T operates on the entire set. To reflect the order of analyses chosen by the investigators, we may write

$$\psi\left(\bigcup_{i=1}^{n} \omega_i\right) = \sum_{i=1}^{n} \psi\left(B_{[i]}\right),$$

and $n_s^{-1} \sum_{s=1}^{n_s} \sum_{\omega_i \subset A} f(\omega_i) \left[\dfrac{\psi\left(B_{[i],s}\right)}{\sum_{\omega_i \subset A} \psi\left(B_{[i],s}\right)}\right]$, where n_s is the number of sequences that were examined by the investigators.

Chapter Summary

In this chapter, we have observed that (1) the mathematical representation of the investigator's choice of their sequence of analyses

---

* The general linear model assessment of subgroups could also provide mathematical justification for this assertion. However, the demonstration is analysis dependent. The set theoretic perspective dismisses the post total cohort assessment subgroup analysis regardless of the perspective.

is an important determinant of the contribution of each analysis to the measure of the union of the collection of analyses and (2) providing the value of $\psi(B_i)$ for each analysis measures the contribution of each analysis's quanta to the union of the collection.

Investigators working within the customary design paradigm make decisions about the priority of analyses. These decisions are based on accuracy, precision, and the persuasive power of the endpoint but also include type I error considerations (i.e., how much alpha should be allocated to each analysis). In this traditional framework, analyses that are not included in the alpha-spending function make no formal contribution to the overall trial assessment.

The measure theoretic infrastructure requires a selected sequence of analyses that is based only on accuracy, precision, and persuasive force. This opens the door to the inclusion of many different analyses to be included in the assessment of the overall research effort.

We are now ready to consider the *topside function* $f(\omega_i)$.

References

[1] Yusuf, S., J. Wittes, J. Probstfield, and H. A. Tyroler. "Analysis and Interpretation of Treatment Effects in Subgroups of Patients in Randomized Clinical Trials." *JAMA* 266, no. 1 (July 3, 1991): 93–8.

Topside Functions

The work that we have invested thus far in this project has been to essentially develop, through an application of measure theory, a "weighting factor" for a collection of functions. Essentially, we have developed every concept and justified each variable in the formula

$$\frac{\int_{\omega_i \subset A} f(\omega_i)\,d\psi}{\int_{\omega_i \subset A} d\psi} = n_s^{-1} \sum_{s=1}^{n_s} \sum_{\omega_i \subset A} f(\omega_i) \left[\frac{\psi(B_{[i],s})}{\sum_{\omega_i \subset A} \psi(B_{[i],s})} \right]$$

except for $f(\omega_i)$. It is now time to develop this and related functions.

As always in our development, we are working with the familiar: our analysis space and σ-algebra (Ω, Σ). This means that any function in the above formula $f(\omega_i)$ must be measurable against (Ω, Σ), which is another way to state that it must (1) not be negative and (2) must derive its value from a property of the analysis ω_i that is available for inspection, as previously discussed. This property will be the plausible interval of the estimator that is available from an inspection of analysis ω_i.

Recall that we began this book with a discussion of duality, i.e., the concept that a single result (be it a serum sodium level in a clinical setting or a research-based, placebo-adjusted difference in exercise tolerance) could simultaneously stand for the occurrence of benefit and of harm.

In this chapter, we return to this idea, now developing the mathematics around it. Once we have completed these new functions' ontogenies, we will combine then with the set theoretic development thus far. Mathematically, we will develop $f(\omega_i)$ for benefit and for harm. We will then accumulate it using

$$\frac{\int_{\omega_i \subset A} f(\omega_i) d\psi}{\int_{\omega_i \subset A} d\psi} = \sum_{\omega_i \subset A} f(\omega_i) \left[\frac{\psi(B_{[i],s})}{\sum_{\omega_i \subset A} \psi(B_{[i],s})} \right].$$

Finally, we will norm this overall analysis paths to compute

$$n_s^{-1} \sum_{s=1}^{n_s} \sum_{\omega_i \subset A} f(\omega_i) \left[\frac{\psi(B_{[i],s})}{\sum_{\omega_i \subset A} \psi(B_{[i],s})} \right].$$

to compute a normed value of this function over all analysis sequences.

This family of functions $f(\omega_i)$ I will call the "topside functions" because they are "topside" or numerator to be divided by the quanta component $\dfrac{n_s}{\psi(B_{i,s})} \sum_{\omega_i \subset A} \psi(B_{i,s})$.

Return to Duality

Remember that we defined duality as the situation where either a lab value or a single estimator of an exposure's effect in a clinical trial can simultaneously support the concept of benefit and the finding of harm.

Statistical hypothesis testing, it will be remembered, simply rejects a null hypothesis, or it does not; the test statistic simply falls into the critical region, or it does not. It is dichotomous in its mathematics.

Duality is a more complex but realistic assessment of what the state of the result is that is wholly consistent with the interpretation in health care.

As we saw, while clinical investigators can quite naturally be flummoxed by the indirect reasoning of dichotomous statistical hypothesis testing, duality is a concept that reflects their experience in interpretation laboratory and imaging results.

Consider an example where the investigators conduct a clinical trial where the primary outcome is the difference in the change of left ventricular ejection fraction between a group exposed to a new intervention and those in the control group. They determine that this placebo-adjusted change in left ventricular ejection fraction (EF) is 6 with a 95% confidence interval of from –2 to 14 based on the standard error.

The standard statistical treatment suggests that the mean placebo-adjusted change in EF observed in this sample is consistent with a population change of zero; the result is therefore declared statistically insignificant.

Thus, dualism is not addressed in statistical hypothesis testing. In that realm, results are reduced to "no difference," "statistically significant increase," or "statistically significant decrease." The idea that the data can support an increase and a decrease simultaneously is lost.

Duality simply allows us to consider this interval of values as an expanse that simultaneously provides evidence of benefit and of harm. In this case, the range of values consistent with harm is quite small (–2–0), while the range consistent with benefit is quite large (0–14).

Interval Parsing, Channeling, and Accumulating

What we have just performed is interval parsing. We started with an interval (in this case, the standard 95% confidence interval, a concept that we will soon expand) and from that interval, plucked an interval of values consistent with benefit (0–14) and similarly an interval of values consistent with harm (0–2). This is the parsing component.

What we will do next is channel the benefit interval through a function (0–14) that assesses and norms this interval so that it is unitless. We will perform the analogous operation for the harm interval (–2–0).

We then repeat this process for all analyses that are responsive to the clinical research question, accumulating these results for the benefit regions. We repeat this process for the harm region and then compare the two. It is this accumulation that is accomplished by the formula

$$n_s^{-1} \sum_{s=1}^{n_s} \sum_{\omega_i \subset A} f(\omega_i) \left[\frac{\psi(B_{[i],s})}{\sum_{\omega_i \subset A} \psi(B_{[i],s})} \right].$$

Once complete, we will carry out an analogous collection and integration for each analysis's interval region of harm. In the end, we will compare the two.

The duality is incorporated by parsing the plausible intervals of benefit and channeling that region into a benefit function and then conducting the analogous operation for harm.

Our Initial Concerns

Recall from chapter 1 that we had two initial concerns about this approach. One was the relationship between the variables being evaluated, i.e., the issue of correlation. This is a straightforward adjustment, but we have not specifically covered this yet.*

However, we have specifically addressed the second issue of the commonality of observations and variables across analyses. The development of ψ-measure is our way of taking into account that many of the same observations and variables are common to these analyses. Therefore, accumulating the impact of the universe of analyses conducted in the study (our ultimate goal) has to, in some way, adjust for the multiple use of observations and variables.

The concept of ψ-measure tells us exactly how to discount analyses for the previous use of their observations and variables. This was, if you will, the hard part. All we need to do is to develop the topside function by (1) build the plausible interval, (2) parse that interval, and (3) describe the benefit and harm functions through which these parsed intervals are to be channeled.

Beginning Construction of the Plausible Interval

Assume that question q concerns the benefit or harm of an intervention in a clinical trial, e.g., "Does the provision of mesenchymal cells to patients with heart failure ameliorate their signs and symptoms when compared to the experience of controls?" Let's describe the set of analyses that address question q as $A_q = \{\omega_i \, / \, q_i \, @ \, q\}$.

---

* It is discussed in a later chapter.

For each analysis ω_i that is a member of A_q, we identify an estimate of effect. For example, it can be the difference between therapy groups of a mean proteomic measure change over time or the relative risk of death associated with the intervention.

Then for each of these estimators, we consider the distorting role of sampling error, bias, and imprecision on this estimate.

Thus, producing the plausible interval for ω_i begins with an inspection of ω_i. One of its elements is the effect size produced by the analysis; this effect size we will notate as e_i. This quantity e_i is any legitimate and well-recognized statistical estimator.

Setting the Bounds for the Plausible Interval

Clinical trial analyses produce statistical estimators of an exposure's effect. Investigators, epidemiologists, and statisticians all recognized that a sample-based effect estimate, being a single number, is influenced by factors unrelated to the effect of the intervention being tested. These factors introduce uncertainty into the effect's true location.

One such type of variability is that generated by taking a sample. This variability is measured by the variance, or its related quantities, the standard deviation, and standard error.[*] Although taken from the same population, different samples of that population contain different patients with different life experiences. Thus, estimates of an effect vary from sample to sample. It is this variability around which the 95% confidence interval is constructed.

However this is not the only reason that the sample value of an estimate differs from the population value. One of these other influences is bias. Bias is a systematic influence on the location of the effect-size estimator. There are many biases in clinical research.[†] Fortunately, the degree to which these biases affect a research result can be identified from a detailed examination of the research design. The identification of a particular bias can aid in determining if the observed effect size is too low

---

[*] The standard error is the standard deviation of an estimate, such as a sample mean or a regression parameter estimate.

[†] Selection bias, recall bias, ascertainment bias, misclassification bias, and immortal time bias are but several of a plethora of biases that can influence an effect-size estimator.

or too high, allowing one to change the bounds of the plausible interval accordingly.

An additional influence is imprecision. Imprecision is the degree to which the measuring instrument provides a different estimate of an individual's data measurement when the measurement is taken repeatedly. For example, Butler et al.[1] report that repeat measurements of left ventricular ejection fraction using the same methods in the same patients by experts in echocardiography routinely vary by 7%. Imprecision is separate and apart from sampling variability, and may not be incorporated into 95% confidence interval.

Bias, imprecision, and sampling variability together blur the true location of the effect-size estimator, injecting uncertainty into its actual value.

This blurring of the effect size's location suggests that both larger values and smaller values of the estimator are admissible for consideration. This range of values will be termed the estimator's interval of plausible effects. It is not just the estimator that provides a sense of the effect of the intervention; it is the estimator's interval of plausible values that is most informative about the possible effect size that would be seen in the population.

The principal reason to name our new interval a plausible interval is to differentiate it from the concept of a confidence interval. The confidence interval is based only on sampling error. The plausible interval includes both the influence of bias and the effect of imprecision as well.

Thus, the plausible interval is a refraction of the effect location based on the particulars of the individual study design, measurement instrument characteristics, and sampling error. It will therefore be wider than the confidence interval because it includes the additional factors of imprecision and bias.

Here, we will develop this concept generally, leaving actual examples to the following chapter.

Define the upper e_i^+ and lower e_i^- bounds of the plausible interval of an estimator from an analysis ω_i and compute

$$e_i^+ = e_i + a_i$$
$$e_i^- = e_i - b_i \, ,$$

where a_i and b_i are constants based on bias, imprecision, and variability. Note that this interval need not be symmetric around the actual estimator e_i. The interval of plausible effect is signified as $\left[e_i^-, e_i^+\right]$.

Parsing the Plausible Interval

This plausible effect interval is to be parsed into two subintervals— one a region of benefit and the other of harm. In order to locate these subintervals, knowledge of the value of the statistical estimator's effect that is neutral (i.e., denotes neither benefit nor harm) is required. Define this value of neutral effect as $e_i(0)$. Similarly, let $e_i(b)$ and $e_i(h)$ be the values of the worst possible benefit and the greatest possible harm permitted by the estimator respectively.[*] Using this notation, then the interval $\left[\min(e_i(h), e_i(b)), \max(e_i(h), e_i(b))\right]$ is the universe of possible values of the estimate.

To compute the plausible interval, consider the case where the greater the benefit, the greater the value of the estimator $e_i(b) > e_i(h)$.[†] We now define the plausible benefit interval $\chi_i^{(b)}$ as

$$\chi_i^{(b)} = \left[b_i^-, b_i^+\right]$$
$$= \left[e_i^-, e_i^+\right] \mathrm{I} \left[\min(e_i(0), e_i(b)), \max(e_i(0), e_i(b))\right]$$
$$= 1_{\left[e_i^-, e_i^+\right]} 1_{\left[e_i(0), e_i(b)\right]} = 1_{\left[b_i^-, b_i^+\right]}.$$

This is the portion of the plausible effect-size region that supports benefit. As an example, consider left ventricular ejection fraction. Larger values of left ventricular ejection fraction are considered beneficial, ceteris

---

[*] The introduction of $e_i(b)$ and $e_i(h)$ is necessary since values of harm need not always be less than values of benefit. For example, if the ith analysis is a total mortality hazard function analysis, then $e_i = 1$ indicates no effect on the time to death $e_i(h) = -\infty$, and $e_i(b) = \infty$. Alternatively, if ω_i is an evaluation of changes in mean differences where the greater differences are salubrious, then the value of $e_i = 0$ reflects no mean effect $e_i(h) = -\infty$, and $e_i(b) = \infty$.

[†] Analagous development is available for the circumstance for estimators, e.g., relative risks, where commonly, the larger the value of the estimator, the greater the harm.

paribus; its increases are beneficial, and its decreases are harmful. Thus, if the plausible effect region, for a change, in left ventricular ejection fraction is $[-2,7]$ and the region of these changes that are beneficial is $(e_i(0), e_i(b)) = (0, \infty)$, then $\chi_k^{(b)} = [-2,7] \cap (0, \infty) = (0, 7]$ is the plausible benefit region.

The plausible interval for harm is based on $\left(\min(e_i(h), e_i(0)), \max(e_i(h), e_i(0))\right) = (-\infty, 0)$, and is

$$\chi_i^{(h)} = \left[h_i^-, h_i^+\right]$$
$$= \left[e_i^-, e_i^+\right] \mathbf{I}\left[e_i(h), e_i(0)\right] = \mathbf{1}_{\left[e_i^-, e_i^+\right]} \mathbf{1}_{\left[e_i(h), e_i(0)\right]} = \mathbf{1}_{\left[h_i^-, h_i^+\right]}$$

which, in this example, is $\chi_k^{(h)} = [-2, 7] \cap (-\infty, 0) = (-2, 0]$.

We will now construct a function from this parsing of the plausibility function into a plausible interval of benefit and a plausible interval of harm.

What Would We Like from a Benefit Function?

If a benefit function is to be compelling, it should increase with increasing benefit and have that benefit be modulated by the presence of uncertainty. If the plausible interval of benefit does not include the null value for effect,[*] the benefit function should be amplified.

Uncertainly is based on the width of the plausible interval of benefit. The greater the width, the less certain we are of the location of the benefit and the less convincing the effect-size estimate is. In addition, we need the benefit function to be unitless so that it can be easily combined with the benefit functions of other analyses for other analyses $\omega_i \subset A_q = \{\omega_i / q_i \text{ @ } q\}$.

These three features are reflected in the function

$$\mathbf{Y}\left(\chi_i^{(b)}\right) = \mathbf{Y}\left(\mathbf{1}_{\left[b_i^-, b_i^+\right]}\right) = r \left[\frac{b_i^- + \dfrac{b_i^+ + b_i^-}{2}}{\left(b_i^+ - b_i^-\right)}\right] e^{-\rho\left(b_i^+ - b_i^-\right)}.$$

---

[*] Recall that the null value is the value for which there is no effect, e.g., 1 for a relative risk or 0 for a mean difference.

Weighing the Evidence

This function maps the interval of plausible benefit to an assessment of the level of that benefit. Benefit is increased when the plausible interval benefit does not include the null effect value. However, the exponential function discounts the benefit by the benefit interval's length $b_i^+ - b_i^-$. The separate component in the denominator, $b_i^+ - b_i^-$, makes the denominator unitless. The parameter r is the proportion of the benefit function that makes up the entire plausible interval. Thus, $Y(\chi_i^{(b)})$ penalizes the benefit estimate derived from ω_i for a wide interval (figure 1).

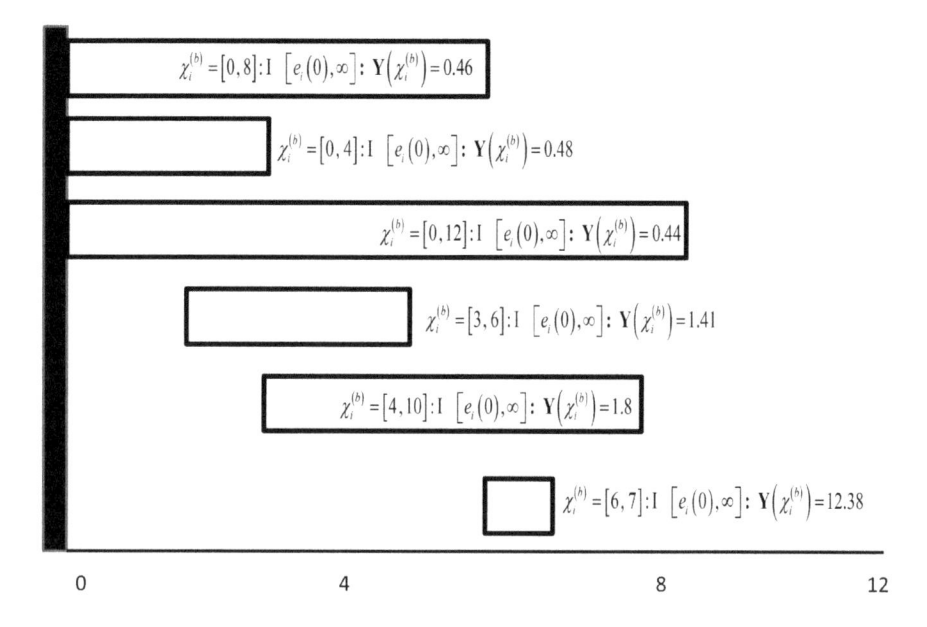

$\chi_i^{(b)} = [0,8]: I \ [e_i(0), \infty]: Y(\chi_i^{(b)}) = 0.46$

$\chi_i^{(b)} = [0,4]: I \ [e_i(0), \infty]: Y(\chi_i^{(b)}) = 0.48$

$\chi_i^{(b)} = [0,12]: I \ [e_i(0), \infty]: Y(\chi_i^{(b)}) = 0.44$

$\chi_i^{(b)} = [3,6]: I \ [e_i(0), \infty]: Y(\chi_i^{(b)}) = 1.41$

$\chi_i^{(b)} = [4,10]: I \ [e_i(0), \infty]: Y(\chi_i^{(b)}) = 1.8$

$\chi_i^{(b)} = [6,7]: I \ [e_i(0), \infty]: Y(\chi_i^{(b)}) = 12.38$

0   4   8   12

Figure 1. Operation of the benefit function for different benefit plausible regions (r=1)

From figure 1 is a circumstance where $\chi_i^{(b)} = [b_i^-, b_i^+] = [0,8]$, $\chi_i^{(b)} = [b_i^-, b_i^+] = [0,4]$ and $\chi_i^{(b)} = [b_i^-, b_i^+] = [0,12]$ each generate a benefit function value of $Y(\chi_i^{(b)})$ of less than one, reflecting some benefit from this region but penalizing this strength because their lower bound includes $e_i(0)$, the value of no effect. The benefit function's value is greater when $b_i^- > e_i^0$, as is the case of the remaining three examples in figure 1.*

A similar quantity can be computed to assess harm. With the plausible harm interval $\chi_k^{(h)}$ defined as above, define

$$\mathbf{Y}\left(\chi_i^{(h)}\right) = (1-r)\left|\frac{1}{\left(h_i^+ - h_i^-\right)}\left(\frac{h_i^+ + h_i^-}{2} + h_i^+\right)\right|e^{-\rho\left|h_i^+ - h_i^-\right|}.$$

Measurable Functions of Benefit and Harm

The notion of benefit and harm can be expanded to an estimate of the size of benefit and the size of harm.

Recall that the plausible benefit interval $\chi_i^{(b)}$ is defined as $1_{\left[b_i^-, b_i^+\right]}$. There are several functions that provide service in assessing the effect of therapy based on that interval. Let **I** be the condition where an increase in e_i reflects benefit and **D** reflect the circumstance where a decrease reflects benefit. Then one such function is $\mathbf{L}_{\max}\left(\chi_k^{(b)}\right) = \mathbf{L}_{\inf}\left(\chi_k^{(b)}\right)\mathbf{1}_\mathbf{D} + \mathbf{L}_{\sup}\left(\chi_k^{(b)}\right)\mathbf{1}_\mathbf{I}$. This measurable represents the assessment of greatest benefit from the plausible interval. It is measurable with respect to (Ω, Σ).

Alternatively, one could conservatively estimate benefit as $\mathbf{L}_{\min}\left(\chi_k^{(b)}\right) = \mathbf{L}_{\sup}\left(\chi_k^{(b)}\right)\mathbf{1}_\mathbf{D} + \mathbf{L}_{\inf}\left(\chi_k^{(b)}\right)\mathbf{1}_\mathbf{I}$. This serves as an extremely conservative estimate of benefit.

Analogously, $\mathbf{L}_{\max}\left(\chi_k^{(h)}\right) = \mathbf{L}_{\sup}\left(\chi_k^{(h)}\right)\mathbf{1}_\mathbf{D} + \mathbf{L}_{\inf}\left(\chi_k^{(h)}\right)\mathbf{1}_\mathbf{I}$ is the worst-case estimate of harm obtained from $\chi_k^{(h)}$, which is obtained from the plausible intervals of harm. We will norm this by dividing by the standard error of the estimate.†

It is without question that there are other estimates of benefits and harm functions available as discussed in the "Limitations" chapter. The

---

* In reality, it is unlikely that the plausible interval for benefit will not include the null value of benefit due to the cumulative effects of sampling error and precision.

† An alternative is to divide by the width of the plausible region. See comments in the limitations chapter.

ones selected here have the advantage of having the required features and are easy to construct.

Now let's put it all together.

References

[1] Butler, J., S. Anker, and M. Packer. "Redefining Heart Failure with a Reduced Ejection Fraction." *Journal of American Medical Association* 322, 18: 761–762.

Putting It All Together

Recall that our purpose for this entire development was to create an analysis platform tailored explicitly (if not exclusively) to clinical research analysis. The required features of this approach were that it (1) include all well-designed analyses (without the need for type 1 error considerations) to address a question posed by the clinical investigators and (2) provide omnibus estimates of effect and harm to address questions in which clinical researchers have the most interest.

The quanta analysis approach, incorporating the concept of duality, is presented as having met these criteria.

We can consider this new method as one of evidence gathering. The clinical researcher, by parsing the plausible intervals of each effect-size estimator using duality theory, is identifying all the evidence from the collection of analyses that support the thesis that the intervention was beneficial.

Performing the analogous evaluation for harm assembles all the evidence from the germane analyses for harm. This evidence is then accumulated and weighed using ψ-measure, which precisely addresses the question of analysis redundancy.

This measure permits us to integrate much more flexibly than statistical hypothesis testing does because it both combines the results of many analyses into an omnibus effect and (through ψ-measure) precisely addresses the complications of analysis redundancy (i.e., that different analyses can substantially overlap in the number of observations and the number of variables used in these analyses).

Mathematically, this approach is translated to the following: for each analysis in the set of analyses responsive to the investigator's question q, i.e., $\omega_i \subset A_q = \{\omega_i / q_i \ @ \ q\}$ investigators identify the benefit functions $\{\mathbf{Y}(\chi_i^{(b)})\}$ and the harm functions $\{\mathbf{Y}(\chi_i^{(h)})\}$. The collection of benefit function results $\{\mathbf{Y}(\chi_i^{(b)})\}$ can now be accumulated over all the analyses $\omega_i \subset A_q = \{\omega_i / q_i \ @ \ q\}$, producing $\int_{A_q} \mathbf{Y}(\chi_i^{(b)})$. Since the analyses are conducted on overlapping sets of observations and variables, the integral is accumulated with respect to ψ- measure and

$$\int_{A_q} \mathbf{Y}(\chi_i^{(b)}) = \int_{A_q} \mathbf{Y}(\chi_i^{(b)}) d\psi.$$

But recall from our earlier discussions that since the contribution of each analysis ω_i to this integral is based on the sequence of analyses (i.e., the paths) that the investigators take through the analyses, we developed the summary integral over all possible paths. Thus, we defined integral of a function f with respect of ψ-measure as

$$\int_{A_q} f(\omega_i) d\psi = n_s^{-1} \sum_{s=1}^{n_s} \sum_{\omega_i \subset A_q} f(\omega_i) \left[\frac{\psi(B_{[i],s})}{\sum_{\omega_i \subset A} \psi(B_{[i],s})} \right].$$ We only have to substitute $\mathbf{Y}(\chi_i^{(b)})$ for f in the above expression to see that the integrated measure of benefit $\mathbf{B}_q$, is

$$\mathbf{B}_q = n_s^{-1} \sum_{s=1}^{n_s} \sum_{\omega_i \subset A_q} \mathbf{Y}(\chi_i^{(b)}) \left[\frac{\psi(B_{[i],s})}{\sum_{\omega_i \subset A} \psi(B_{[i],s})} \right].$$

Similarly, the integrated summary of harm, denoted as $\mathbf{H}_q$, can be written as

$$\mathbf{H}_q = n_s^{-1} \sum_{s=1}^{n_s} \sum_{\omega_i \subset A_q} \mathbf{Y}(\chi_i^{(h)}) \left[\frac{\psi(B_{[i],s})}{\sum_{\omega_i \subset A} \psi(B_{[i],s})} \right].$$

These computations beg the calculation of the benefit-harm ratio as

$$\text{BHR}_q = \frac{\mathbf{B}_q}{\mathbf{H}_q} = \frac{\int_{A_q} Y\left(\chi_i^{(b)}\right) d\psi}{\int_{A_q} Y\left(\chi_i^{(h)}\right) d\psi},$$

which can be expressed as

$$= \frac{n_s^{-1} \sum_{s=1}^{n_s} \sum_{\omega_i \subset A_q} Y\left(\chi_i^{(b)}\right) \left[\frac{\psi\left(B_{[i],s}\right)}{\sum_{\omega_i \subset A} \psi\left(B_{[i],s}\right)} \right]}{n_s^{-1} \sum_{s=1}^{n_s} \sum_{\omega_i \subset A_q} Y\left(\chi_i^{(b)}\right) \left[\frac{\psi\left(B_{[i],s}\right)}{\sum_{\omega_i \subset A} \psi\left(B_{[i],s}\right)} \right]} = \frac{\sum_{s=1}^{n_s} \sum_{\omega_i \subset A_q} Y\left(\chi_i^{(b)}\right) \psi\left(B_{[i],s}\right)}{\sum_{s=1}^{n_s} \sum_{\omega_i \subset A_q} Y\left(\chi_i^{(b)}\right) \psi\left(B_{[i],s}\right)}$$

BHR_q is the benefit-harm ratio obtained from a consideration of all the analyses that the investigators deemed responsive to question q. It is a nonnegative valued and ranged from zero to infinity. Values less than one suggest more harm than benefit, while values greater than one reflect the reverse.

Recall also that we identified an estimate of the least beneficial effect for the intervention based on analysis ω_i, $\mathbf{L}_{\max}\left(\chi_k^{(b)}\right) = \mathbf{L}_{\inf}\left(\chi_k^{(b)}\right) \mathbf{1}_{\mathbf{D}} + \mathbf{L}_{\sup}\left(\chi_k^{(b)}\right) \mathbf{1}_{\mathbf{I}}$, where the plausible benefit interval $\chi_i^{(b)}$ is defined as $\mathbf{1}_{\left[b_i^-, b_i^+\right]}$, $\mathbf{I}$ being the condition where an increase in e_i reflects benefit and $\mathbf{D}$ reflecting the circumstance where a decrease reflects benefit. This function can also be integrated over the entire collection of analyses responsive to question q $\omega_i \subset A_q = \{\omega_i / q_i \,@\, q\}$ as $\Lambda_{q\mathbf{B}}$, where

$$\Lambda_{q\mathbf{B}} = \int_{A_q} \mathbf{L}_{\max}\left(\chi_k^{(b)}\right) d\psi = n_s^{-1} \sum_{s=1}^{n_s} \sum_{\omega_i \subset A_q} \mathbf{L}_{\max}\left(\chi_k^{(b)}\right) \left[\frac{\psi\left(B_{[i],s}\right)}{\sum_{\omega_i \subset A} \psi\left(B_{[i],s}\right)} \right].$$

We can think of Λ_{qB} as the normed beneficial effect of the therapy across all analyses responsive to question q_r. Analogously, we can compute the normed harmful effect across studies, Λ_{qH}, as

$$\Lambda_{qH} = \int_{A_q} \mathbf{L}_{\max}\left(\chi_k^{(h)}\right) d\psi = n_s^{-1} \sum_{s=1}^{n_s} \sum_{\omega_i \subset A_q} \mathbf{L}_{\max}\left(\chi_k^{(h)}\right) \left[\frac{\psi\left(B_{[i],s}\right)}{\sum_{\omega_i \subset A} \psi\left(B_{[i],s}\right)} \right].$$

where, as developed in the previous chapter, $\mathbf{L}_{\max}\left(\chi_k^{(h)}\right) = \mathbf{L}_{\sup}\left(\chi_k^{(h)}\right) \mathbf{1}_D + \mathbf{L}_{\inf}\left(\chi_k^{(h)}\right) \mathbf{1}_I$ is the worst-case estimate of harm obtained from $\chi_k^{(h)}$, which is obtained from the plausible intervals of harm.

The benefit-harm ratio ($\mathbf{BHR}_q$), the normed beneficial effect (Λ_{qB}), and the normed effect of harm (Λ_{qH},) are the major products of this work. Together, they permit investigators to determine the strength of the evidence for benefit and harm and compare them using a single value, reflecting the integrated finding across all relevant analyses. Similarly, Λ_{qB} and Λ_{qH}, provide estimates of the best benefit and worst harm that could be anticipated from exposure to the intervention in the population.

Let's now look at some examples. We will start from the most simple demonstration of quanta theory and then move to scenarios of increasing complication and controversy. Each of these examples are scenarios that occur in health care research.

In each of the following examples, we assume that the results are research efforts from randomized clinical trials that are prospectively designed and concordantly executed and that the prospectively declared outcome data have been obtained as precisely as possible. The purpose of these examples is to provide some calibration to duality theory and then to demonstrate how duality theory is combined with quanta analysis in the circumstance of multiple outcomes.

Example 1: One and Only One Outcome—No Effect Size

This is the simplest of examples that demonstrates the data summarization that is provided by duality analysis.

Weighing the Evidence

In this circumstance, we have a randomized clinical trial that is assessing the effect of a medication on left ventricular ejection fraction (LVEF). The data shown are the results of sixty hypothetical patients randomized to the medication being tested or to a control therapy. Patients have LVEF measured at baseline and at follow-up. The difference in the change in LVEF between the groups is provided as the effect size.

Table 1. Standard vs. duality analysis; single outcome - zero effect size

Standard Result		Duality Analysis	
Effect size	0	Benefit=	0.22
Standard Error of Effect Size	4	Harm=	-0.22
LB for Plausible Effect Region	-12	**B/H Ratio=**	1.00
UB for Plausible Effet Region	12	Best Benefit=	3.00
		Least Benefit=	0.00
		Least Harm =	0.00
		Greatest Harm=	-3.00

In this case, the difference in the change in LVEF across the two groups is 0; this estimate's standard error is 4. The plausible effect region (considering both standard error and imprecision) is ± 12 absolute EF units.

Using the standard statistical paradigm, the conclusion of this study with its one outcome is that there is no effect of therapy on LVEF.

However, the plausible region (that incorporates the standard error) suggests that there may, in fact, be a small positive effect in some patients and a small harmful effect in other patients.* Duality analysis, which considers variability as potential benefit and potential harm, reflects this more nuanced perspective.

---

* The plausible region reflects bias (which, in this case, is zero), imprecision, and the sample variability.

With only a single analysis, there is no analysis rotation required (i.e., there is only one analysis path)[*]; thus, this experiment permits us to examine the correspondence between the effect of therapy and duality as reflected in the topside function of the previous chapter.

On a scale of $(-\infty, \infty)$, where positive values reflect benefit and negative values reflect harm, the benefit function's value in this example is 0.22 and the value of harm is -0.22. They are identical since the plausible interval definition is symmetric and the effect size is 0. Not surprisingly, the benefit-harm ratio is one because in this case of one outcome, it reduces to the ratio of the benefit and the harm estimates.

The estimate of best benefit is the extreme value of the plausible interval associated with benefit, normed by the standard error. The least benefit estimate is the smallest value of the plausible interval of benefit, also normed. Analogous values of harm are provided. So we could conclude from this first assessment that both the standard analysis and the quanta analysis support the same conclusion—the absence of evidence that benefit exceeds harm.

The duality analysis comes to the same conclusion, but with a different emphasis.

Example 2: One Outcome—Moderate Effect Size

Now staying with the same research design, let's inject some benefit into the finding of this experimental scenario by increasing the estimate of the difference in the change in EF from 0 to 7 (table 2). This result produces a standard test statistic of $7/4 = 1.75$, which would not meet the 0.05 threshold.

The quanta analysis provides a different assessment. Note that the benefit-harm ratio has increased from 1.00 in the previous example to 3.30. The conclusion from this one analysis is that the evidence for benefit is greater than the evidence for harm.

---

[*] Our assessments of benefit and harm are based on $\int_{A_s} f(\omega_i) d\psi = n_s^{-1} \sum_{s=1}^{n_s} \sum_{\omega_s \subset A_s} L_{max}(\chi_k^{(h)}) \left[\dfrac{\psi(B_{[1],s})}{\sum_{\omega_s \subset A} \psi(B_{[1],s})} \right]$, which, in the case of only one analysis, reduces to $\int_{A_s} f(\omega_1) d\psi = f(\omega_1)$.

This inclination for benefit is also reflected by the best benefit and greatest harm estimates. Note in example 1 that the finding of zero effect produce symmetric estimates of best benefit and least harm. In table 2, that symmetry in the duality analysis disappears.

Table 2. Standard vs. duality analysis; single outcome - moderate effect size

Standard Result		Duality Analysis	
Effect size	7	Benefit=	0.33
Standard Error of Effect Size	4	Harm=	-0.10
LB for Plausible Effect Region	-5	**B/H Ratio=**	3.30
UB for Plausible Effet Region	19	Best Benefit =	4.75
		Least Benefit =	0.00
		Least Harm =	0.00
		Greatest Harm =	-1.25

In the duality analysis, the finding for benefit is greater than that for harm (in absolute value). Thus, the benefit-harm ratio has increased from the value of 1 in table 1 to 3.30 in table 2. A standard statistical hypothesis test on the table 2 findings would simply conclude there was no benefit.

However, the duality analysis concludes that while there is evidence of benefit and harm, there is more evidence of benefit than of harm. The duality analysis also quantifies the best benefit (as a unitless number) of 4.75 and the greatest harm as −1.25.

Note that in each of these two examples, the least benefit and greatest harm are each zero. This is because the plausible interval crosses zero, the location of null effect. Since the finding is for benefit, the estimate of best benefit is greater than the finding for greatest harm (in absolute value).

Circumstances where the plausible interval does not cross the null value, suggesting that there is no evidence that any subject in the study experienced any harm (on the germane outcome measure), is quite rare and is not a circumstance developed in any of the presented analyses in the book.

Example 3: One Outcome—Large Effect Size

In a further modification of this exercise, in example 3, we increase the effect size from 4 to 9. As we might anticipate at this point, the finding for best benefit has increased with the effect size, and the finding for greatest harm has moved closer to zero, producing a benefit-harm ratio of 18.46 (table 3).

Table 3. Standard vs. duality analysis; single outcome - large effect size

Standard Result		Duality Analysis	
Effect size	9	Benefit=	0.38
Standard Error of Effect Size	4	Harm=	-0.02
LB for Plausible Effect Region	-5	**B/H Ratio=**	**18.46**
UB for Plausible Effet Region	19	Best Benefit =	5.75
		Least Benefit =	0.00
		Least Harm =	0.00
		Greatest Harm =	-0.25

Example 4: One Outcome—Overwhelming Harm Effect

In order to demonstrate that these functions work as expected for a finding of harm, an example where the effect size is now negative is provided.

Table 4. Standard vs. duality analysis; single outcome - finding for harm

Standard Result		Duality Analysis	
Effect size	-5	Benefit=	0.14
Standard Error of Effect Size	4	Harm=	-0.30
LB for Plausible Effect Region	-17	**B/H Ratio=**	**0.46**
UB for Plausible Effet Region	7	Best Benefit =	1.75
		Least Benefit =	0.00
		Least Harm =	0.00
		Greatest Harm =	-4.25

In this circumstance, the difference in the change in LVEF is now negative. Here, the duality analysis reveals there is more evidence for harm than evidence for benefit. The benefit-harm ratio is now less than one. In addition, the greatest harm estimate is now –4.25, and the best benefit estimate is only 1.75, also demonstrating a shift to harm.

Conclusions from Single Outcome Examples

The scenarios of the previous examples in this were to simply provide some calibration for the operation and results of duality analysis in basic clinical trial analyses with a single outcome.

They are not the basis of a supplantation argument; one would not replace statistical hypothesis testing based on these simple comparisons.

We saw that when there was clear benefit, the duality functions demonstrated benefit. They also appropriately revealed that harm was present when the standard statistical estimator suggested harm. When the effect was zero, they demonstrated essentially equivalent benefit and harm.

These findings are what we would expect.

Example 5: Two Outcomes with Reversed Effects

The challenge now to be faced is how the combination of duality theory and the application of measure theory through quanta analysis functions operate when there is more than one outcome.

Example 5 examines one of these more complex scenarios. In this prospectively designed, concordantly executed clinical trial, there are two outcomes: the difference in the change in LVEF and the difference in the change in LV end systolic volume (ESV). Neither one of these predominates from a pathophysiologic perspective; there is no natural primary outcome here.

The traditional, statistical, hypothesis-testing domain requires that either one be prospectively chosen as a primary outcome or that type I error be distributed between the two. A duality or quanta evaluation does not require these actions.

In this circumstance, we have chosen to permit the outcomes to not provide results that demonstrate the same direction of clinical effect (table 5). The difference in the change in ejection fraction of 7 demonstrates a change in the direction of benefit, yet the change in end systolic volume is in a harmful direction.*

---

*   Increases in end systolic volume, ceteris paribus, are harmful.

Table 5. Standard vs. quanta analysis; two outcomes in discrepant directions

Standard Result	EF	ESV	Duality Analysis	EF	ESV
Effect size	9	25	Benefit=	0.35	0.07
Standard Error of Effect Size	4	12	Harm=	-0.06	-0.23
LB for Plausible Effect Region	-3	-11	**B/H Ratio=**	**5.85**	**0.32**
UB for Plausible Effet Region	21	61	Best Benefit =	5.25	0.92
			Least Benefit =	0.00	0.00
			Least Harm =	0.00	0.00
			Greatest Harm =	-0.75	-5.08

The duality analysis, as we might expect from the previous four examples, recapitulates the standard result; substantially more benefit than harm is seen for the EF outcome, while harm is predominant for the ESV outcome.

Statistical hypothesis testing is of limited use here. If the LVEF outcome was declared as primary, then the study would be viewed from a statistical perspective as positive, a finding that would be contradicted by the ESV findings. However, if the ESV outcome were declared as primary, the study would be viewed as demonstrating harm, a result that would ignore the important LVEF finding. If the type I error had been divided in two, with one half apportioned to each of the two outcomes, then neither outcome would be seen as statistically significant and equally unsatisfactory to clinical researchers because that result contradicts the data.

Thus, it would be the artificial process of selecting one or two primary outcomes, and not the data themselves, that determine the study's findings using the traditional statistical hypothesis testing approach. Yet neither of these conclusions is correct or satisfactory.

Duality theory tells us that while benefit and harm are evident in each of the two outcomes, the magnitude of benefit and harm differ across the two outcomes. We now turn to quanta analysis to accumulate these findings across the two endpoints. This approach permits to integrate the findings of benefit and harm over the two outcomes and assemble a final number reflecting this combined assessment.

Recall that since there is more than one outcome, the finding that we take depends on the path or rotation of analysis that is used in the integration. To consider all possibilities, we integrate over all possible paths (table 6).

Table 6. Rotation Summary

Sequence of En	EF	ESV	Summary
1 2	180(.6)	120(.4)	
2 1	120(.4)	180(.6)	
Rotation	1, 2	2, 1	
Benefit=	0.24	0.19	0.21
Harm=	-0.13	-0.16	-0.15
B/H Ratio=	**1.89**	**1.15**	1.52
Best Benefit =	3.52	2.65	3.08
Least Benefit =	0.00	0.00	0.00
Least Harm =	0.00	0.00	0.00
Greatest Harm =	-2.48	-3.35	-2.92

Table 6 demonstrates this path analysis (for orientation, in the rotation, 1 denotes EF, and 2 is ESV). With only two endpoints, there are only two possible rotations or paths: 1, 2; or 2, 1.

The first two lines of table 6 provides the weights or ψ-measures for each analysis in their sequence. For example, $\psi(B_{LVEF}) = 180$, when LVEF is considered first, and $\psi(B_{ESV|LVEF}) = 120$, which is the quantum contribution of the ESV analysis after considering the measure of the EF analysis.

When EF is considered first on the analysis path, its analysis consumes 60% of the total available measure, with ESV making up the additional 40%. The next line provides $\psi(B_{ESV})$ and $\psi(B_{LVEF|ESV})$, the required quantities when the ESV outcome is the first in the path.

The subsequent data in the table provide the results of the integration as function of the rotation sequence or paths. In each circumstance, both the beneficial findings of the intervention on LVEF and the harmful effects on ESV are considered. However, their contributions depend on the quanta weights. In the first column, the beneficial effect of EF predominates because it is based on 60% of the available measure. However, the **BHR** (benefit-harm ratio) is not as high as that seen in the duality analysis for EF since it is combined with the finding from ESV.

Similarly, the second column of table 6 reveals that the harmful effect observed with ESV is modulated by the beneficial effect of EF.

The final summary column averages the duality theory findings over both paths. Note that the benefit-harm ratios are substantially modulated from 5.85, observed when LVEF was considered alone and the (0.32) when ESV was considered by itself (table 5). Overall, both benefit and harm are seen with a benefit-directional effect. The best and worst benefit and harm cases are provided as well.

In this case, statistical hypothesis testing is at best inconclusive. The duality or quanta analysis demonstrates that there is a modest but overall benefit when considering both LVEF and ESV findings simultaneously.

Example 6: Three Outcomes Each with Small Effects

In the next scenario, we consider the circumstance of clinical trial with three outcomes: LVEF, ESV, and LV end diastolic volume (EDV). In this scenario, there is a modest increase in LVEF and with modest decreases in each of ESV and EDV. The standard statistical paradigm suggests that this is a "negative study" (table 7).

Table 7. Standard vs. quanta analysis; three outcomes - all "negative"

Standard Result				Duality Analysis			
	EF	ESV	EDV		EF	ESV	EDV
Effect size	7	-17	-38	Benefit=	0.33	0.35	0.31
Standard Error of Effect Size	4	12	25	Harm=	-0.10	-0.11	-0.09
LB for Plausible Effect Region	-5	-53	-113	**B/H Ratio=**	**3.30**	**3.21**	**3.66**
UB for Plausible Effet Region	21	61	37	Best Benefit=	4.75	4.42	4.52
				Least Benefit=	0.00	0.00	0.00
				Least Harm=	0.00	0.00	0.00
				Greatest Harm=	-1.25	-1.58	-1.48

Each of these outcomes reveals that there is an inclination to benefit; however, in the standard statistical paradigm, each of these findings would fail the 0.05 p-value criteria, much more so with any correction for multiplicity.

The duality evaluation for each of these endpoints, with its recognition that each of these findings reflects some evidence for benefit and for harm, reveals that the benefit-hazard ratio supports benefit for each of the findings.

Weighing the Evidence

The quanta analysis is more complicated because there are six different paths or sequences to be considered in the assembly of the overall finding. It integrates over the six different paths and then summarizes (table 8).

Table 8. Rotation Summary

Outcome Path	EF	ESV	EDV				
1 2 3	180(.43)	120(.29)	120(.29)				
1 3 2	180(.43)	120(.29)	120(.29)				
2 1 3	120(.29)	180(.43)	120(.29)				
2 3 1	120(.29)	180(.43)	120(.29)				
3 2 1	120(.29)	120(.29)	180(.43)				
3 1 2	120(.29)	120(.29)	180(.43)				

Rotation	1, 2, 3	1, 3, 2	2, 1, 3	2, 3, 1	3, 2, 1	3, 1, 2	Summary
Benefit=	0.33	0.33	0.33	0.33	0.33	0.33	0.33
Harm=	-0.10	-0.10	-0.10	-0.10	-0.10	-0.10	-0.10
B/H Ratio=	**3.36**	**3.36**	**3.35**	**3.35**	**3.41**	**3.41**	3.37
Best Benefit=	4.59	4.59	4.54	4.54	4.56	4.56	4.56
Least Benefit=	0.00	0.00	0.00	0.00	0.00	0.00	0.00
Least Harm=	0.00	0.00	0.00	0.00	0.00	0.00	0.00
Worst Harm=	-1.41	-1.41	-1.46	-1.46	-1.44	-1.44	-1.44

Table 8 demonstrates the findings for each of the six possible paths in the integration. The entries in the upper half of the table demonstrate the quanta weights for each of the endpoints in each of the six paths (1 = LVEF, 2 = ESV, 3 = EDV). The second half of the table demonstrates the result of the integration for each of the six paths and then a summary measure. It is supportive of benefit, a finding that is not surprising since each of the outcomes demonstrated benefit.

Note that each of the six rotations essentially provided the same result—benefit directionality. This is principally because each outcome demonstrated more benefit than harm. Thus, even though the quanta weights differed from rotation to rotation, for each rotation evaluation, there was a finding of benefit, and therefore, the summary finding across all rotations is one of benefit.

Example 7: Three Outcomes with One Disparity

However, suppose one of the outcomes did demonstrate a singular, disparate effect (table 9).

Table 9. Standard vs. quanta analysis; three outcomes - one "positive"

Standard Result	EF	ESV	EDV	Duality Analysis	EF	ESV	EDV
Effect size	4	-20	23	Benefit=	0.28	0.42	0.17
Standard Error of Effect Size	4	8	25	Harm=	-0.15	-0.04	-0.12
LB for Plausible Effect Region	-8	-44	-52	B/H Ratio=	1.85	10.57	1.36
UB for Plausible Effet Region	16	4	98	Best Benefit=	4.00	5.50	2.08
				Least Benefit=	0.00	0.00	0.00
				Least Harm=	0.00	0.00	0.00
				Greatest Harm=	-2.00	-0.50	-3.92

In this case, the results for LVEF trends to benefit (i.e., LVEF moves in the direction of improvement but do not reach statistical significance). The decrease in ESV is pronounced and also statistically significant. However, the EDV finding moves in the direction of harm.

This circumstance is problematic for clinical investigators who receive little help from statistical hypothesis testing, a tool too inflexible to manage this complicated set of results coherently and helpfully.

The duality analysis demonstrates the strong ESV finding as a large benefit-hazard ratio of 10.57. The **BHR**$_q$ ratios of the other two endpoints are quite small. However, the findings of the quanta evaluations that accumulate the duality analysis results from each outcome are unambiguous (table 10).

Table 10. Rotation Summary

Outcome Path	EF	ESV	EDV
1 2 3	180(.43)	120(.29)	120(.29)
1 3 2	180(.43)	120(.29)	120(.29)
2 1 3	120(.29)	180(.43)	120(.29)
2 3 1	120(.29)	180(.43)	120(.29)
3 2 1	120(.29)	120(.29)	180(.43)
3 1 2	120(.29)	120(.29)	180(.43)

Rotation	1, 2, 3	1, 3, 2	2, 1, 3	2, 3, 1	3, 2, 1	3, 1, 2	Summary
Benefit=	0.29	0.29	0.31	0.31	0.27	0.27	0.29
Harm=	-0.11	-0.11	-0.10	-0.10	-0.11	-0.11	-0.11
B/H Ratio=	**2.58**	**2.58**	**3.23**	**3.23**	**2.53**	**2.53**	**2.78**
Best Benefit=	3.88	3.88	4.09	4.09	3.61	3.61	3.86
Least Benefit=	0.00	0.00	0.00	0.00	0.00	0.00	0.00
Least Harm=	0.00	0.00	0.00	0.00	0.00	0.00	0.00
Worst Harm=	-2.12	-2.12	-1.91	-1.91	-2.39	-2.39	-2.14

The first half of table 10 is the same as table 9 since the number of observations and variable have not changed.* The bottom half of table 10 shows the findings for each rotation. Note that rotations 2, 1, 3 and 2, 3, 1, since they begin with ESV, provide the greatest **BHR**$_q$ for the overall effect. However, even for these two paths, the **BHR**$_q$ is modulated since it is combined with the weaker findings of benefit from LVEF and EDV.

The overall finding of benefit is reduced from the isolated finding of 10.57 to more moderate findings of 2.78. This is a reasonable finding based on the data.

Statistical hypothesis testing, with its requirement to select a single path of analysis (in this case, a primary and two secondary endpoints), was much too restrictive since the investigator did not know which single outcome to select as primary.

The quanta analysis avoids this by considering each endpoint in turn as a primary (or first in analysis), each as second, and then each as third; and combining the finding.

With this experience, we can now examine a new take on some outstanding issues in clinical research efforts.

---

*   The quanta computation is based on not just the path of analyses but also on the number of common observations and variables between the outcomes, and since this has not changed, the quanta are the same.

Quanta Analyses and the Supremacy of Safety

"Primum non nocere" is the supreme principle that governs the practice of physicians. It determines their role in the delivery of medical services to individual patients and sets in place the relationship between health care research and its volunteer subjects.

In classic clinical trial analyses, safety is embedded in the protocol design. It is one of the reasons that we have control groups in clinical trials.* It is a rationale for two-sided statistical hypothesis testing in clinical research and is one of the principal motivations for interim analyses. Each of the prospectively declared outcomes in a clinical trial is assessed for harm as well as for benefit.

Safety is a principal function of the Data and Safety Monitoring Board (DSMB), one of whose tasks is to monitor the experience of all subjects in a clinical trial and to react to safety threats by recommending protocol changing, including discontinuing the study.

Institutional review boards are a mainstay of local oversight of the conduct of a clinical trials. The United States federal Food and Drug Administration (FDA) devotes substantial resources to the review of safety data from medication and devices approved in the United States, both prior to and after the intervention's approval.

---

* One of the arguments that Bradford Hill made to his colleagues in the first clinical trial involving streptomycin was that the presence of a control group would help attribute antibiotic-induced adverse effects appropriately to the antibiotic, allowing them to discontinue it if it was unsafe.

This all makes sense from public health and epidemiologic perspectives. And certainly, the morality and ethics of placing safety first are beyond reproach.

My purpose here is not to criticize health research workers in their approach to safety in their investigative endeavors but to instead demonstrate how the philosophy of safety preeminence can play a more central and quantitative role in summarizing the findings of their research efforts.

The Safety Disconnect in Research

The safety experience of clinical research participants predominates in clinical research and contributes to the balanced interpretation of the study. However, there is something of a disconnect in reporting findings from clinical research.

Commonly, the efficacy findings of a research program are detached from the safety findings. For example, efficacy and safety findings appear in different portions of the results section of a manuscript, with efficacy findings appearing earlier than those of safety. There is no attempt to assimilate the two in the results section; this occurs noetically in the discussion section.

While clinical researchers have become accustomed to this, there is no scientific reason why this must be so, especially since both safety and efficacy must ultimately be integrated in the end, if only cerebrally, to have an assessment of risk and benefit.

However, the quantitative separation between efficacy findings and safety findings is convenient because there are no widely used statistical methods through which this integration can take place.

Safety Findings and Type I Error

By the current standard, family-wise type I error, so carefully parsed for the collection of prospectively declared efficacy endpoints, is not included in the assessment of safety. In general, safety testing is at the nominal 0.05 alpha level. There are typically no corrections for multiplicity for safety evaluations. Essentially type I error for safety endpoints is treated much like that for secondary efficacy endpoints.

Weighing the Evidence

This traditional approach is not without its own rationale.* Sharing the type I error apportioned for efficacy with safety analyses decreases the type I error for each assessment. The results of hypothesis testing regarding safety outcomes would need to be more extreme in order to meet the lower alpha level. In addition, strictly requiring the type I error level control for safety findings is not consistent with a Safety First philosophy.

In addition, we must keep in mind that primary efficacy endpoints are assessed for harm as well as for efficacy, an assessment that occurs under full type I error control, so it is not as though all safety evaluations across therapy groups are nominally interpreted.

Finally, the remaining lower alpha level for the efficacy endpoints would increase the sample size of the study considerably because error rates, just like event rates and efficacy, are powerful drivers of the numbers of patients required for the study. This is wholly due to a type I error and not a safety consideration.

What Else Can We Do?

These rationales are fine, but they do not overcome the argument for a formal and quantitative integration of safety and efficacy findings in clinical research efforts.

The actual explanation for the absence of such integration is that we have no reliable way in biostatistics (and certainly no reliable way using statistical hypothesis testing) to conduct this integration. Since safety is an important occurrence in the use of the study, then a quantitative synthesis of safety and efficacy findings could be useful. However, while desired, it has been unavailable.

However, our duality and quanta analysis approach can accommodate this request. From its perspective in the accumulation of analyses responsive to question q, it does not matter whether the analyses that are to be integrated are efficacy analyses or safety analyses, i.e., the set of analyses $\omega_i \subset A_q = \{\omega_i / q_i \quad q\}$ can include safety analyses as well. We simply need to determine the analysis path.

---

* The argument that safety outcomes are so important that they transcend type I error considerations is vacated by the observation that statistical hypothesis testing is used to assess differences in safety endpoints across therapy groups.

In addition, the use of the duality principle allows us to assess the evidence of benefit and evidence of harm from the safety outcomes, just as for efficacy, accumulating them using the exact same procedures that we developed for the their accumulation of efficacy and harm from efficacy outcomes. Parsing, channeling, and accumulating works as easily for safety as it does for efficacy. Thus, these safety analyses are passed through the benefit and harm functions $\mathbf{Y}\left(\chi_i^{(b)}\right)$, and $\mathbf{Y}\left(\chi_i^{(h)}\right)$ and are gathered up with these functions that are evaluated for efficacy analysis in $\int_{A_q} \mathbf{Y}\left(\chi_i^{(b)}\right) d\psi$ and $\int_{A_q} \mathbf{Y}\left(\chi_i^{(h)}\right) d\psi$ respectively.* No new theory is required; we simply need to expand the same processes of parsing, channeling, and accumulation into and through the safety assessments.

Example—Heart Failure Therapy and Creatinine

For example, consider a clinical trial designed to assess the effect of a new therapy for heart failure. This randomized and double-blinded clinical trial will assess the effect of the therapy on well-established outcomes, e.g., left ventricular ejection fraction (EF), left ventricular end systolic volume (ESV), and left ventricular end diastolic volume (EDV). However, since the intervention is anticipated to have a nephrotoxic effect, the difference in the change in creatinine levels across the two groups is also measured.

The results, in accordance with the standard analysis and the duality analysis, reveal more evidence for benefit than harm with the three efficacy outcomes; however, as expected, there is an increase in the mean creatinine level associated with therapy (table 1).

---

* This is also true for the computation of the estimate of benefit Λ_{qB} and harm Λ_{qH},

Table 1. Standard vs. duality analysis quanta analysis; three primary and 1 safety outcome

Standard Result

	EF	ESV	EDV	Cr
Effect size	4	-20	-35	0.2
Standard Error of Effect Size	4	12	25	0.07
LB for Plausible Effect Region	-8	-56	-110	-0.01
UB for Plausible Effet Region	16	16	40	0.41

Duality Analysis

	EF	ESV	EDV	Cr
Benefit=	0.28	0.36	0.31	0.01
Harm=	-0.15	-0.09	-0.09	-0.49
B/H Ratio=	**1.85**	**3.79**	**3.50**	**0.02**
Best Benefit=	4.00	4.67	4.40	0.14
Least Benefit=	0.00	0.00	0.00	0.00
Least Harm=	0.00	0.00	0.00	0.00
Greatest Harm=	-2.00	-1.33	-1.60	-5.86

However, we see that unlike the three efficacy evaluations, there is greater evidence for harm than benefit for changes in serum creatinine levels. This is where the analysis typically ends in the standard paradigm, which is confirmed by the duality analysis.

However, we are in a position to ask, "What does the integrated result look like?" In order to answer this, the one remaining question that we must answer is where in the path the safety analysis should appear—if we place it either before (Safety First) or after (Safety Last) the three efficacy analyses (table 2).

Table 2. Impact of Location of Safety Analysis on Summary Findings

	Safety Last	Safety First
Benefit=	0.25	0.21
Harm=	-0.20	-0.24
B/H Ratio=	**1.28**	**0.91**
Best Benefit=	3.42	2.95
Least Benefit=	0.00	0.00
Least Harm=	0.00	0.00
Greatest Harm=	-2.58	-3.05

Table 2 provides the rotation summary. In the first column, we have the result by including safety last in the path. The implication of this last position is that this safety analysis has the smallest ψ-measure. However, even there, we see that there has been an important reduction in the benefit-risk ratio despite the benefit findings from the three efficacy endpoints visible from table 1.

However, when safety is considered first in the quanta analysis path, there is a marked decrease in the benefit-hazard ratio even further, tipping it to harm.

Summary

Thus, not only can safety evaluations be folded into an integrated summary of the result of the study but it is also possible to consider their impact first before consideration of the impact of the primary endpoints, which is wholly consistent with a "safety first" mentality. Such a procedure will likely change the assessment of some clinical trials.

Managing Correlation Between Variables

The development of duality theory and quantum analysis has thus far been silent on the issue of correlation. As clinical investigators understand, while it is reasonable to assume that the individual participants of a clinical research effort make independent contributions to an analysis, the variables utilized in an analysis are commonly correlated among themselves. It stands to reason that this dependency will affect the contribution of a collection of variables to the ψ-measure of an analysis.

Recall that the ψ-measure of an analysis ω_i is simply $\psi(\omega_i) = n_i v_i$, where n_i is the number of observations and v_i the number of variables used in ω_i. In a single analysis, it is possible that v_i could be large. This, of course, would substantially increase the ψ-measure.

However, if these variables in the analysis of ω_i are correlated, then they are (colloquially expressed) "measuring the same thing." This redundancy in what they measure should decrease the measure of the analysis. This thought process is the motivation for taking a reduction action on the ψ-measure of this analysis by taking the correlation among its variables into account.

This is a technical chapter. For those who are willing to accept that correlation can be incorporated into ψ-measure, then please proceed to the chapter on exploratory evaluations. Those who have a background in multivariable analysis in general and determinants in particular or are interested in seeing its possible application to ψ-measure, please feel free to charge ahead through this chapter's development.

Proposed Formulation Using Determinants

A formulation for the measure of an analysis that incorporates this dependency is $\psi(\omega_i) = n_i v_i |\mathbf{R}(\underline{v}_i)|$, where $|\mathbf{R}(\underline{v}_i)|$ is the determinant* of the v_i by v_i correlation matrix of all the variables used in the analysis ω_i.

In order to get a sense for the mechanics of this formulation, consider that when $v_i = 2$, $\psi(\omega_i) = n_i v_i |\mathbf{R}(v_i)| = n_i 2 \begin{vmatrix} 1 & r \\ r & 1 \end{vmatrix} = n_i 2(1 - r^2)$.

Thus, when the correlation is minimal, then each of the two variables substantially contribute to the analysis's measure. When the dependency is high, the contribution of the two variables to the measure of the analysis diminishes due to fact that to some degree, these two variables measure the same phenomenon.

Correlations and Unions of Analyses

Recall that the ψ-measure of the union of two analyses ω_i and ω_j is $\psi(\omega_i \cup \omega_j) = \psi(\omega_i) + \psi(\omega_j) - \psi(\omega_i \cap \omega_j)$. In order to formulate the role of correlation within the measure of the union of analyses, we must consider how it is embedded in the measure of their intersection.

Our consideration of correlation should reflect not just the correlation of variables used in the same analysis but also the more *distant* correlation of variables that are different between the two analyses, yet nevertheless correlated with each other. This latter circumstance should permit these "inter-analysis" correlations to contribute to the measure of the overlap between the analyses.

We can proceed by asserting that if $\psi(\omega_i) = n_i v_i$, and $\psi(\omega_j) = n_j v_j$, then we can assemble the measure of the intersection

---

* The determinant is computed from a matrix of numbers. It is a measure of the degree to which linear combinations of some of the columns can be used to reproduce other columns. When the matrix is not just numbers but is instead the correlation coefficients between variables (i.e., a correlation matrix), then the determinant of that correlation matrix is a reflection of redundancy in the system. In a system of n variables, the largest the determination can be is 1, reflecting no dependency, and the smallest the determinant can be is zero, reflecting complete redundancy (i.e., at least one of the variables can be reproduced by adding multiples of some of the other variables).

$$\psi_{ijk...m} = \psi\left(\omega_i \cap \omega_j \cap \omega_k \cap ... \cap \omega_m\right) = n_{ij}v_{ij}\left(d + |\mathbf{R}_u|\right).$$

Here, d is the number of variables that are in common between the two analyses, and u is the number of variables that are not in common but may be correlated. This permits correlated variables that are not common across the analyses to contribute to the measure of the intersection through their correlations.

The following are examples of how these formulations operate.

Example 1—Regression Analysis Families

Consider the circumstance in which a collection of straight-line regression analyses are carried out in a clinical research effort. The dependent variable is the same for each analysis—the change in left ventricular ejection fraction from baseline to follow-up. This change is the response variable that is assessed against a baseline value of 1 of m phenotypes.

Thus, in this collection of analyses, there are m different regression analyses, each a function of three variables (two variables determine the change value, and 1 variable is the phenotype). Each analysis has the same number of participants n. The goal is to compute the measure of $\psi\left(\bigcup_{i=1}^{m}\omega_i\right)$. As is our process, define $\{B_j\}$ as the collection of disjoint sets, such that $\bigcup_{i=1}^{m}B_i = \bigcup_{i=1}^{m}\omega_i$, and $\psi\left(\bigcup_{i=1}^{m}\omega_i\right) = \sum_{i=1}^{m}\psi(B_i)$.

Let's first presume that the phenotype variables are independent of one another. In this circumstance, calculation for $\psi(B_j)$, $j=1...m$ is straightforward because $\psi\left(\bigcap_{i=1}^{m}\omega_i\right) = 2n$ (each analysis contains the same number of participants and has two variables in common). Thus,

$\psi(B_1) = 3n.$

$\psi(B_2) = 3n - 2n = n.$

$\psi(B_3) = 3n - 2n - 2n + 2n = n.$

$\psi(B_4) = 4n - 2n - 2n - 2n + 2n + 2n + 2n - 2n = n.$

...

In general, $\psi(B_j) = 3n1_{j=1} + n1_{1<j\le m}$ and

$$\psi\left(\bigcup_{i=1}^{m} \omega_i\right) = \sum_{i=1}^{m} \psi(B_i) = 3n + (m-1)(n) = n(m+2).$$

However, in reality, the m phenotype variables are correlated. The implication of this correlation is that the measure of any intersection among a collection of the $\omega_i, i = 1...m$ analyses written now as

$$\psi\left(\bigcap_{i=1}^{m} \omega_i\right) = 2n$$ will be a misrepresentation of the intersection's measure.

Following the development of the methods section for dependence, write

$$\psi(B_2) = \psi(\omega_2) - \psi(\omega_1 \cap \omega_2)$$
$$= 3n - n(d + (1 - |\mathbf{R}_{12}|)) = 3n - n(2 + (1 - |\mathbf{R}_{12}|)).$$

The measure of the intersection of the first two analyses ω_1 and ω_2 $\psi(\omega_1 \cap \omega_2) = n(d + (1 - |\mathbf{R}_{12}|))$ reflects the observation that (1) the same number of n participants are included in each regression analysis and (2) the number of variables in common across the two analyses is $d + (1 - |\mathbf{R}_{12}|)$, where $d = 2$ reflects that in each analysis, the same two variables determine the responder analysis. $|\mathbf{R}_{12}|$ is the determinant of the 2 × 2 correlation matrix for phenotypes 1 and 2. Since

$$|\mathbf{R}_{12}| = \begin{vmatrix} 1 & r_{12} \\ r_{12} & 1 \end{vmatrix} = 1 - r_{12}^2, \text{ find}$$

$$\psi(\omega_1 \cap \omega_2) = n(d + (1 - (1 - r_{12}^2))) = n(2 + r_{12}^2).$$

When the correlation is zero, then the number of common variables is 2, and the solution defaults to the earlier case in this example. However, as the correlation increases, the greater is the propensity of two phenotype variables to measure the same feature, and the greater their contribution is to the measure of the two analyses' intersection. Thus,

$$\psi(B_2) = \psi(\omega_2) - \psi(\omega_1 \cap \omega_2) = 3n - n(2 + r_{12}^2) = n(1 - r_{12}^2).$$

Note that if $r_{12} = 1$, then $\psi(B_2) = 0$. This follows from the observation that if the correlation is 1, then the two phenotypes essentially measure the same relationship, and the incremental value of the regression on the second phenotype is zero after assessing the first regression.

Continuing on, find

$$\psi(B_3) = \psi(\omega_3) - \psi(\omega_1 \cap \omega_3) - \psi(\omega_2 \cap \omega_3) + \psi(\omega_1 \cap \omega_2 \cap \omega_3)$$
$$= 3n - n(2 + r_{13}^2) - n(2 + r_{23}^2)$$
$$+ n(2 + (1 - |\mathbf{R}_{12}|) + (1 - |\mathbf{R}_{13}|) + (1 - |\mathbf{R}_{23}|))$$
$$= 3n - n(2 + r_{13}^2) - n(2 + r_{23}^2) + n(2 + r_{12}^2 + r_{13}^2 + r_{23}^2)$$
$$= n(1 + r_{12}^2).$$

The calculation proceeds for $i > 3$.

Summary

The procedure laid out in this chapter is certainly not the only way to incorporate correlation into quanta analysis. However, it is consistent with what ψ-measure is attempting to assess, is relatively easy to compute, and is smoothly integrated into the calculation of the quanta measures.

Incorporating Exploratory Analyses

The flexibility of the duality and quanta analyses lies principally in the quanta component. The implication of ψ-measure means that any collection of analyses that can be assembled from our set theory tools of unions, intersections, and complements can be measured.

This is the principal advantage of requiring ψ to be a measure. Thus, it can assign weight to combinations of safety and efficacy endpoints, measure the various paths through efficacy outcomes, and permit us to disconnect from the notion of declaring a prospectively described outcome as a primary one simply from statistical considerations.*

However, ψ-measure also provides freedom that we may not know how to handle. This brings us to exploratory analyses.

Exactly What Are Exploratory Analyses?

Exploratory analyses are evaluations conducted in health care research that were not prospectively declared.† Commonly described

---

* That is, for the assignment of type I error prospectively

† There is very little written about the theory of exploratory analyses beyond statements about the problems such analyses can cause, e.g., Nicenboim, B., S. Vasishth, F. Engelmann, and K. Suckow. "Exploratory and Confirmatory Analyses in Sentence Processing: A Case Study of Number Interference in German." *Cogn Sci* 42, S4 (June 2018):1075–1100, https://onlinelibrary.wiley.com/doi/10.1111/cogs.12589.

Much of this section is from my article: Moyé, L. "What Can We Do about Exploratory Analyses in Clinical Trials?" *Contemp Clin Trials* (Sep 18, 2015), https://doi.org/10.1016/j.cct.2015.09.012. Epub ahead of print.

as retrospective or post hoc appraisals, these assessments follow no prospective plan and have no protocol.

The motivation for the analyses might be findings identified in other studies or a de novo observation from the investigators' own ongoing research. Because these post hoc evaluations appear to answer questions that the investigator did not think to ask prospectively, exploratory analyses can be engaging and even exciting.

There are many reasons to conduct unplanned, hypothesis-generating analyses in clinical trials. In some circumstances, during the course of a study, a new outcome measure is determined to be of value in a second trial. With this new outcome in hand, the research community is particularly interested in how that new outcome measure performs in the current trial. The investigators may, in fact, be compelled to report this outcome even though it was not part of the prospective plan.

In other circumstances, a journal reviewer or the editor may desire to see a particular analysis. The reason is this is typically well motivated, and the clinical investigators, anxious to satisfy these arbiters of publication, will provide the analysis, which may or may not be published.

In fact, it is not uncommon for the United States federal Food and Drug Administration (FDA), when reviewing the voluminous filings of pharmaceutical companies that are supplied in support of a product, to ask for additional analyses from pivotal phase III studies that were not prospectively declared by the investigators. Of course, these requests are granted. The National Institutes of Health also engages in these analyses.[1]

Also, there is the investigational motivation to examine all the data at hand in new combinations to see if a new facet of the disease can be examined.

The use of exploratory analyses is not surprising. Unanticipated findings play an undeniable role in science. Radiation was not found because it was sought but because the researchers stumbled across it. Minoxidil and sildenafil are examples of medications that were designed for one purpose, but perspicacious investigators identified unanticipated new effects that opened the door to new indications.[*]

It is therefore no surprise that exploratory analyses are critical in first-in-human studies. In these cases, there is very little information to

---

[*] Both minoxidil and sildenafil were developed as antihypertensives. Minoxidil was discovered to be one of the first effective medical treatments for alopecia, and sildenafil evolved into a treatment for erectile dysfunction.

determine the universe of effects of a biologic agent or small molecule. Thus, these early protocols tend to be especially restrictive. The observed effects generated by a biologic or small molecule—by the very nature of the poor state of a priori knowledge—are likely to be a surprise.

An example is determination of optimum cell preparations for cell therapy clinical trials.[2] There is particular interest in publishing exploratory analyses in oncology, with the exploratory components clearly marked.[3]

Therefore, phase I studies have and should have a heavy exploratory component since the earlier the study, the less is known and the more is identified post hoc. There is little question that such findings must be confirmed in a standard confirmatory evaluation. However, we must also acknowledge that if the question was not first raised by the exploratory finding, the confirmation would not have been forthcoming.

The Problem with Exploratory Analyses

However, there are difficulties with exploratory analyses. The principal difficulty is not with the data evaluation itself but with the interpretation of the data.

Early clinical trial experience did not differentiate prospectively declared evaluations from nonprospectively declared assessments. If the p-value was less than 0.05, the result was considered not just statistically significant but also valid and reliable.

In chapter 2, we saw the problems with this approach. The MRFIT program, as well as INVEST, ELITE, PRAISE, and the US carvedilol programs are just some of the examples that demonstrated that there were hazards with nonprospectively declared outcomes. The source of these hazards is practical and theoretical.

> *Logistical Concerns*
> The principal justification given for the unreliability of an exploratory analysis is the effect of the absence of prospective planning on the precision of the exploratory estimators of effect size. We will call this the *logistical rationale*.[4]
>
> For example, if investigators wish to conduct a clinical trial on the effect of an intervention on heart

muscle perfusion, they are obligated to ensure superior quality and high precision images for the endpoint measures, e.g., identifying a core laboratory. These trial design controls reduce endpoint variability and, ceteris paribus, increase power.

However, should these same investigators observe at the study's conclusion a treatment attributable benefit for coronary artery disease death, they will be hard-pressed to defend the reliability of this unanticipated finding. The absence of its prospective declaration meant that there was no opportunity to organize resources for its reliable estimation.

For example, without prior definition of coronary artery disease death, there could be no a priori structure in place for the formal collection of death records and no endpoint committee of specialists to adjudicate findings. In addition, the analysis suffered from an absence of prospective statistical planning that, had it been present, would have produced both informative power computations and the minimum number of deaths required to draw a conclusion with some statistical regularity.

Theoretical Concerns

A second concern is more theoretical. It is the random selection of the analyses. Prospective outcomes are chosen based on knowledge of mechanism of action, reliable data, and the availability of precise outcome estimators, before the data are collected.

Exploratory analyses, on the other hand, are essentially chosen for publication randomly. They are selected due to their unanticipated small p-value. When chosen in this fashion, we have learned that these analyses are not reliable (i.e., they are unable to be reproduced).

Does that Mean They Should Not Be Published?

Exploratory evaluations are commonly the first databased view of the future. Today's exploratory research can be tomorrow's new confirmatory outcome. Thus, exploratory evaluations should have a place in the literature.

They just have to be clearly labeled as exploratory and not permitted to replace the findings of the prospectively declared outcomes.

A problem is that it is difficult to have the results of exploratory analyses published in some areas of research whether they are clearly labeled as such or not. It is as though to the editors, the very moniker *exploratory* is too controversial. The opisthotonic reaction of journals to exploratory analyses that are clearly labeled as requiring replication does a disservice to the medical research community.[*]

The inclusion of exploratory analyses clearly labeled as such raises type I error concerns as well. In addition to concerns about the interpretation of p-values in statistically underpowered environments, one cannot prospectively apportion type I error in non-prespecified environments.[†]

Duality and Quanta Analyses in Exploratory Evaluations

The advantage that quanta analysis holds is that exploratory analyses can be incorporated into the omnibus benefit-risk ratio while discounting their contributions for the reasons outlined above.

To be clear, we will only consider exploratory analyses that are well conducted as worthy of inclusion. Sloppy evaluations, low quality data, and imprecise definitions have no place in any study, whether it be based on statistical hypothesis testing or duality or quanta analyses.

---

[*] There are counter examples to this. For example, there is interest in publishing exploratory analyses in oncology as previously mentioned in this chapter. In addition, there is substantial interest in the behavioral sciences in understanding exploratory factor analysis.

[†] While one could allocate, for example, 15% of the available type I error to exploratory analyses, the impact on the sample size is not inconsiderable and would be judged too big a price to pay to assess analyses unknown to the clinical investigators at the start of the study.

With these exclusions, there are exploratory outcomes that are measured precisely. For example, the variables that are produced from cardiac MR imaging provide volumes of data and variables about which little are known. All the MR variables, be they prospectively declared or not, are measured with the highest possible precision. Consider a clinical trial with two treatment groups and three outcomes—one primary and two secondary. They are the difference in the change in left ventricular ejection fraction (primary), the difference in the change in left ventricular end systolic volume (ESV; secondary), and difference in the change in left ventricular end diastolic volume (EDV; secondary).

The data demonstrates no clinically important change in these outcomes. However, there is one MR based exploratory outcomes measured at high precision: MR1 (table 1).

Table 1. Standard vs. duality analysis quanta analysis; three primary and 1 exploratory outcome

Standard Result

	EF	ESV	EDV	MR1
Effect size	3	-2	5	3.5
Standard Error of Effect Size	4	12	25	2
LB for Plausible Effect Interval	-9	-38	-70	-2.5
UB for Plausible Effect Interval	15	34	80	9.5

Duality Analysis

	EF	ESV	EDV	MR1
Benefit=	0.27	0.24	0.17	0.36
Harm=	-0.17	-0.17	-0.12	-0.10
B/H Ratio=	**1.57**	**1.44**	**1.39**	**3.54**
Best Benefit=	3.75	3.17	2.80	4.75
Least Benefit=	0.00	0.00	0.00	0.00
Least Harm=	0.00	0.00	0.00	0.00
Greatest Harm=	-2.25	-2.83	-3.20	-1.25

Note that the duality analysis demonstrates evidence for more (but not much more) benefit than harm for each of EF, ESV, and EDV. However, it does note substantially more evidence for benefit from the exploratory endpoint. However, for the reasons that have been provided in this chapter, considering the MR exploratory analysis on par with the prospectively declared non-MR analyses is problematic. However, the exploratory evaluation can occur last in the path analysis (table 2).

Table 2. Summary of clincal trial results including exploratory outcome

	Summary
Benefit=	0.26
Harm=	-0.14
B/H Ratio=	1.81
Best Benefit=	3.63
Least Benefit=	0.00
Least Harm=	0.00
Greatest Harm=	-2.37

Only two rotations are required since the primary outcome always appears first, the exploratory outcome is last, and there were only two secondary outcomes. The result reveals that there was a small increase in the benefit-hazard ratio due to the exploratory outcome. However, the uptick was modest given its position as last in the analysis path, where it retained over 22% of the total ψ- measure.

Clearly, the more primary and secondary endpoints there are, ceteris paribus, the less measure is left for the exploratory outcome.

The purpose of this chapter was to demonstrate the flexibility of the duality and quanta approaches through its ability to readily absorb exploratory analyses. However, this facility should not dominate concerns over exploratory analyses. Many such evaluations are not worthy of further consideration due to poor or missing data and incompletely conceived analysis plans. Mathematics cannot adumbrate these grave matters.

References

[1] Kaltenthaler, E., C. Carroll, D. Hill-McManus, A. Scope, M. Holmes, S. Rice, M. Rose, P. Tappenden, and N. Woolacott. "The Use of Exploratory Analyses within the National Institute for Health and Care Excellence Single Technology Appraisal Process: an Evaluation and Qualitative Analysis." *Health Technol Assess* 20, 26 (April 2016): 1–48. https://doi.org/10.3310/hta20260.

[2] Landin, A. M., and J. M. Hare. "The Quest for a Successful Cell-based

Therapeutic Approach for Heart Failure." *Eur Heart J* (Jan 10, 2017). Epub ahead of print. https://doi.org/10.1093/eurheartj/ehw626.

[3] Tahara, M., M. Schlumberger, R. Elisei, M. A. Habra, N. Kiyota, R. Paschke, C. E. Dutcus, T. Hihara, S. McGrath, M. Matijevic, T. Kadowaki, Y. Funahashi, and S. Sherman. "Exploratory Analysis of Biomarkers Associated with Clinical Outcomes from the Study of Lenvatinib in Differentiated Cancer of the Thyroid." *Eur J Cancer* 75 (April 2017): 213–221. Epub Feb 24, 2017. https:/doi.org/10.1016/j.ejca.2017.01.013.

[4] Moyé, L. "What Can We Do about Exploratory Analyses in Clinical Trials?" *Contemp Clin Trials* (September 18,2015). Epub ahead of print. https:/doi.org/10.1016/j.cct.2015.09.012.

Contributions of Other Measurable Functions

The mathematical concept of measurability plays a central role in this book's development. For example, we have measurable functions **Y** and **L** through which the parsed plausible intervals for benefit and harm are channeled. We have also established a formal measure ψ to help us manage precisely the redundancy in analyses, permitting to accumulate these functions over regions of analyses, e.g., $\int_{A_q} \mathbf{Y}\left(\chi_i^{(b)}\right) d\psi$.

However, there are additional possible uses for measurable functions in our application of duality theory to health care research.

Recall that a measurable function must meet three criteria: (1) it must be real-valued, (2) it must be nonnegative, and (3) it must generate its numerical value based on an inspection of the properties of ω_i.

Recall that any particular analyses ω_i from our Ω has many different properties attached to it. Thus far, we have utilized very of few of them (the statistical estimator, the plausible interval, whether it was a safety analysis, whether it was an exploratory analysis), but other properties of the analysis should and can influence the impact of an analysis. Since the use of these can be the basis of a measurable function on (Ω, Σ), we can create helpful measurable functions that permit us to modulate or amplify the influence of a particular analysis on the assessment of benefit and harm.

One such influence is that of the characteristics of an analysis. Setting statistical hypothesis testing aside, there are other features of analyses that we quite correctly consider when assessing the impact of an analysis.

A critical feature is whether the analysis is prospective or retrospective (exploratory). Other features have to do with the presence of a contemporary control group, the presence of randomization, and a degree of blinding.* When these features are present, we provide more weight to the analysis.

How would this work mathematically?

Recall that all of our work have revolved around the concept of $\int_{\omega_i \subset A_q} f(\omega_i) d\psi$, which is an accumulation of the function $f(\omega_i)$ (which, for us, has been a benefit function or a harm function) over all analyses responses to the research q, expressed as $\omega_i \subset A_q = \{\omega_i / q_i \quad q\}$.

If we wish to modify or amplify the function $f(\omega_i)$ without fundamentally changing the function f, we can simply develop another measureable function $\mathbf{M}(\omega_i)$ (M for "methodology"). This function will place the highest value on analyses that have the strongest methodology.

We can implement this function quite simply. One way is to permit the set function $m_j(\omega_i)$ to be the jth methodologic property of analysis ω_i and is assigned a value based on the presence of that property. These properties are quite easily enunciated (table 1).

Table 1. Defining Methodologic Features for $\mathbf{M}(\omega_i)$.		
j	Methodology Characteristic	Value
1	Prospective analysis	3
2	Use of a core lab	2
3	Pilot study control group	1
4	Pilot study randomization	1
5	Pilot study—blinded	1
6	Pivotal study control group	2
7	Pivotal study randomization	2
8	Pivotal study—blinded	2
9	Adequate sample size	2

---

*   These last three features are hallmarks of clinical trials.

Table 1 provides some possible values for the analysis characteristics. For example, if the analysis is prospectively designed from a pilot study, which had a control group but was neither randomized nor blinded, then $\sum_{j=1}^{9} m_j(\omega_i) = 4$.

We can then define

$$\mathbf{M}(\omega_i) = \frac{2e^{\sum_{j=1}^{m} m_j(\omega_i)}}{1 + e^{\sum_{j=1}^{m} m_j(\omega_i)}} - 1.$$

This is a function that is trapped between 0 and 1. For analyses that have relatively weak methodology, $\mathbf{M}(\omega_i)$ is close to zero. Analyses with the strongest methodologies produce values $\mathbf{M}(\omega_i)$ close to one. Our benefit and hazard integrals would then be written as

$$\mathbf{B}_q = \int_{\omega_i \subset A_q} \mathbf{Y}\left(x_i^{(b)}\right) \mathbf{M}(\omega_i) d\psi$$

$$\mathbf{H}_q = \int_{\omega_i \subset A_q} \mathbf{Y}\left(x_i^{(h)}\right) \mathbf{M}(\omega_i) d\psi.$$

The multiplication within the integral permits us to modulate the effect of the benefit or hazard function based on the methodology that is utilized by the analysis. This feature operates independently of the path analysis. An analysis that occurs early on the analysis path but is crippled by its weak methodology (for example, the absence of blinding) will have reduced influence on the benefit and harm interval.

This feature provides the clinical investigator the second of two control features that manage the impact of an analysis.

Limitations

Is duality/quanta analyses ready for prime time?
No.
It is a fine idea, but it has limitations and contains arbitrary decisions that must be more closely examined with the goal of improvement. No doubt you have identified your own such set of concerns. Here is mine.

The Quanta Measure

As iterated in this text, the quanta measure is the key to this tool's flexibility. However, the decision that $\psi(\omega_i) = n_i v_i$ was arbitrary. It was selected because it a simple quantity directly related to the data and when investigated, proved to be a measure and is easily computed.

However, this is not the only such measure; in fact, there are uncountably many measures available. The use of $\psi(\omega_i)$ is useful, but it must be seen as only a starting point in this new field of exploration.

The most important new ingredient is not the specific formulation of $\psi(\omega_i)$, but the use of a formal measure that permits wide latitude in computing the value of a union of analyses. This is the new key component. What the best measure might be is wholly up to debate, discussion, and improvement.

The Topside Function Is Not Optimal

Parsing the plausible interval into one of benefit and one of harm is simple. How one incorporates these fragments into a function is complicated, and I confess that this part of the project took far more time than I anticipated.

Of one thing, I am certain: this topside function can be improved. From my perspective, the most important features that it must have are the following:

1. It provides a unitless measure of benefit and one of harm;

2. The wider the plausible interval, the less emphasis the actual estimator of benefit and harm receive; and

3. The further the lower bounds of the plausible intervals for benefit and harm are from the null value (i.e., that value which indicates no effect), the greater the strength of the finding.

4. The functions for greatest/least benefit/harm can be normed by the width of the plausible interval.

In addition, the assumption that the region of plausible values must include the possibility of benefit and of harm in research (that is, they must cover the null value of the estimate) is valid, I believe, but it requires reexamination.

My experience informs me that in studies that demonstrate even overwhelming benefit, there are individuals who receive the exposure who are harmed by it, either by an outcome (e.g., diastolic blood pressure) moving in the wrong direction or the occurrence of a safety event reliably attributed to the exposure. This was my motivation for assuming the plausible interval must include the null value.

The investigators have full freedom in choosing the plausible intervals, but it is recommended that they be wide, not narrow. Its goal is not to include only probable effects but also those that are unlikely but possible since the improbable commonly occurs in health care.

The Methodology Function Is Not Unique

A measureable function $\mathbf{M}(\omega_i)$ on our usual (Ω, Σ) regions of analysis was introduced. This function was designed as a conduit for the impact of the methodologic rigor of the research effort on the functions of benefit and harm produced by that effort's estimators. $\mathbf{M}(\omega_i)$ was defined as quantitative metric based on the characteristics of the research method. This is clearly not the only definition of such a function. There are other measures of the quality of a research effort. In addition, there are other functional forms besides $\mathbf{M}(\omega_i) = \dfrac{2e^{\sum_{j=1}^{m} m_j(\omega_i)}}{1 + e^{\sum_{j=1}^{m} m_j(\omega_i)}} - 1$. These should be developed and examined.

There Is No Sample Size Formula

This is an important concern. The determination of the number of subjects that there should be in a research effort is a critical, practical consideration for researchers because it is an important driver of the logistics (how many recruiting centers and coinvestigators are required) and the likelihood of funding. Its absence is a crucial impediment to the implementation of duality or quanta analyses as the only evaluation of the contribution of the research effort.

However, this is just a technical issue. Our mathematical development demonstrates that the quanta contribution to our measurement of, for example, benefit from the collection of questions $\omega_i \subset A_q = \{\omega_i / q_i \quad q\}$ is

$$\mathbf{B}_q = \int_{\omega_i \subset A_q} \mathbf{Y}(\chi_i^{(b)}) \mathbf{M}(\omega_i) d\psi$$

$$= n_s^{-1} \sum_{s=1}^{n_s} \sum_{\omega_i \subset A_q} \mathbf{Y}(\chi_i^{(b)}) \mathbf{M}(\omega_i) \left[\dfrac{\psi(B_{[i],s})}{\sum_{\omega_i \subset A} \psi(B_{[i],s})} \right].$$

The quanta component, $\dfrac{\psi(B_{[i],s})}{\sum_{\omega_i \subset A} \psi(B_{[i],s})}$, since it represents a collection of percentages, is sample-size independent as is $\mathbf{M}(\omega_i)$, which is a measure of methodologic rigor. The impact of the sample size is therefore on the contribution of the benefit function, which is itself a function of the plausible interval for benefit. Recall that this plausible interval is a function of accuracy, precision, and bias of the statistical estimator. Thus, we simply need a sample size that controls these three features across each of the estimators in the set $\omega_i \subset A_q = \{\omega_i / q_i \quad q\}$. This must be developed, but it also should be tractable.

Lack of Independent Confirmation

A disadvantage of this development is that it was conducted by one and only one person—the author. Assessments by independent researchers—either in recapitulating my own efforts or through their own derivations, regenerating my conclusions—is essential. The raison d'être for this book is to call for that process while, at the same time, producing my work for this required review.

A Real Work Test Is Lacking

Duality/quanta analysis must be put to the actual test, i.e., actual clinical research data should be run through it. This will improve the robustness of the software and also provide some calibration of the unitless results. I have no access to clinical trial data at this point, so I cannot provide this essential calibration myself. However, this research effort must be put to test with real data.

Conclusions—Queen Anne's Decree

Are we better off with a new process for mathematically assessing the impact of health care research? My answer is yes. I have provided one in this book.

Duality and quanta analysis is not the only alternative, and it may not be the best alternative. But it is not beset with the well-known panoply of weaknesses that afflicts statistical hypothesis testing that were reviewed in this book's earlier germane discussion.

Quanta Analysis

Duality and quanta analysis address the straightforward, central, almost Reaganesque question[*]: "Are my patients better off being exposed to the intervention?" Through a process of parsing, channeling, and accumulating, it gathers and weighs the evidence for and against benefit. This evidence can be accumulated across many analyses in clinical research that bear on the research question.

Duality or quanta analysis respects the role of prospective evaluations while also providing a way to incorporate safety analysis (at the beginning) and exploratory analyses with precise outcomes (at the end) of the analysis paths.

Finally, it provides the basis for additional mathematical research to sharpen its contribution to health care research.

---

[*] In a 1980 US presidential debate, candidate Ronald Reagan reduced the complex cultural and economic questions facing voters by asking "Are you better off now than you were four years ago?"

The limitations of this approach have been provided; however, those limitations are simply methodology and will be removed through continued mathematical developments and with experience that comes from practical use of these tools with real data.

The Need for a Solid Research Foundation Endures

The duality and quanta approaches recommended in this text is an alternative to statistical hypothesis testing, but it is not the only alternative. Over the generations, Bayes procedures and artificial intelligence algorithms consistently show promise, but the community energy is not behind these approaches (or any *p*-value alternative).

However, any new approach requires a solid epidemiologic and logistical foundation. The investigation must be well designed. Logical contemporary control groups, precise endpoints, and sensible outcomes must be selected. Effect sizes must make clinical sense. A quantitative metric of successes should be preannounced.

With this solid runway in place, several different methodologies besides this text's findings can be used to bring the research plane in for a successful landing.

Cultural Conflict of Interest

Reviewing quanta or duality analysis as well as other admissible substitutes as possible *p*-values replacements is an imperative. However, we—biostatisticians and, to some degree, clinical investigators—must begin this review with an acknowledgment of our own conflicts of interests.

Our conflict of interest is not necessarily financial but intellectual. It is a cultural conflict of interest.

Statistical hypothesis testing has been in wide use since the mid-1950s. The decades from then to now span the working careers of most all of us.[*] We understand how to conduct statistical hypothesis and have some comfort level with their degrees of complexity.

For many biostatisticians, work consists of designing research endeavors that will produce *p*-values, discussing which *p*-value is most

---

[*] Mine began in my third year of medical school in 1977.

suitable to the circumstance, and generating tables with *p*-values. Our careers have become *p*-value–centric.

A move away from statistical hypothesis testing holds important implications for the suitability of our knowledge base, our productivity, and our careers. If would be a profound midcourse correction.

Clinical investigators, while in a somewhat different boat, are caught up in the same current. These researchers are forced to develop some facility with *p*-values since these computations are required for grant applications, the consideration of result-laden abstracts at influential meetings, and ultimately manuscript publication.

Thus, clinical investigators have become used to *p*-values even if they do not like them. However, despite their mistrust and (commonly) miscomprehension of them, when a clinical investigator's results are accompanied by small *p*-values, the researchers can't help but rejoice, regardless of the effect size. So, although in general *p*-values in health care research may be rancid butter, they are on the right side of the investigator's bread when less than 0.05.

We therefore must acknowledge that taking a step away from statistical hypothesis testing while introducing opportunity also injects new uncertainty into research programs.

Some will resist any change simply because it is change. This later group has a vested interest in seeing statistical hypothesis testing remain in place and will fight to keep its standing.

Whatever our position in the community, and whatever our perspective, we must manage our conflict.

Taking Matters into our Hands

Clinical investigators can no longer wait to be rescued from statistical hypothesis testing. As we have seen, this patience has so far been rewarded by the straitjacketing of complex, clinical research interpretation and the attempt by some quantitative workers to "double down" on the *p*-value, driving it from 0.05 to 0.005. Since these are not a helpful solution as pointed out earlier, investigators must therefore take matters into their own hands.

They have the power they need to affect change. It is clinical investigators—not biostatisticians—who develop the ideas for new interventions. It is the clinical investigators who understand the disease and the population of patients in whom the intervention will be used. They also understand the necessary duration of follow-up and the serious, adverse events when they occur.

While these critical abilities do not permit investigators to choose the metric that will objectively assess their work, these capabilities empower them to resist the current, statistical, hypothesis-testing metric. They can call for its replacement, play a role in developing the criteria for its replacement, contribute to the choice of a new metric, and monitor its performance.

Clinical investigators can no longer be innocent bystanders being hit by p-value crossfire. We have to fight our way to a place at the table.

Researchers should have at their disposal a wide range of biostatistical support procedures including, but not requiring, statistical hypothesis testing. This flexible approach is wholly consistent with the desire of researchers seventy years ago when all recognized the sad state of affairs of research protocols. It provides the rigor of a well-conceived protocol but is not p-value–centric.

Statistics should be contributory and supportive, but not dominant in research design. That is a position reserved for clinical investigators and epidemiologists.

The role of NIH and the FDA is to support these new innovative efforts, not enforce an obsolete administrative and statistical metric of success simply because it is the system to which they have become accustomed.

These federal administrators have a need for an administrative metric for assessing the results of a research program. This should be developed for them. But they should no more enforce a single metric of statistical data analyses on investigators than university researchers should dictate to NIH the funding pay line the institute uses.

Longitude

Perhaps what is required is a new version of the emergency decree of Queen Anne.

In the late seventeenth century, European maritime commerce was failing. Although the development of the New World was well underway, it could not be reliably reached by even the best ships.

While late arrivals were commonly explained by the ship's captain as "losing the weather gauge"[*] en route, the principal reason for delay was that ships were lost at sea. This occurred because, while seamen could easily determine their latitude[†], they could not find their longitude. The absence of a real-time longitude assessment had grave consequences for trade development as well as important naval ramifications.

In response to this, at the end of Queen Anne's reign, the English Parliament passed the Longitude Act of 1714. It established a Board of Longitude and offered monetary rewards to anyone who could establish a simple and practical method for the determination of a ship's longitude. This generated excitement in the maritime community; many different ideas were suggested, and ultimately a solution was found.[‡]

While apparently many clinical investigators and biostatisticians are "lost at sea" when asked to define a p-value,[§] there is sadly little "wind in the sails" of the statistical community to find a p-value replacement.

However, an Act of Congress could provide sufficient financial and career inducement to the identifier of an interpretable and practical replacement for the p-value. The selection process, to be managed by the National Science Foundation, could help us turn our backs on an obsolete rule, and help clinical investigators to finally find their way home.

---

[*] Losing the weather gauge meant no longer having a favorable wind.

[†] From the date and duration of the day

[‡] Even though this board was rife with conflicts of interest as members demonstrated favoritism for different candidates, in the end, they settled on the best solution. The winning candidate produced an accurate, portable clock that would work reliably at sea. With two such clocks on board, one reading London time and the other the time at high noon on the ship, one could convert the difference in the time to a specific change in longitude.

[§] As stated in the preface, confusion among statisticians became so bad that the American Statistical Association, for the first time in its 177-year history, felt compelled to issue a statement clarifying for its own membership what p-values mean and how they should be used. This statement led to further clarifications.

Biographies

Georg Cantor

The one mathematician above all who is responsible for catapulting set theory from an arcane finite and interesting contrivance to the basis of modern mathematics is Georg Cantor.[1]

He probably died for it.

In the nineteenth century, mathematics had not yet escaped the grasp of religion. One such captive content area was infinity. While all mathematicians knew that the counting numbers were infinite, very little was understood about the concept of infinity. The prevalent intuition was that of eternal life, and since God created that, then that must be where God had a special place for himself. And if the natural numbers were infinite, then must be He who inhabited them.

The oldest of six children, Georg Cantor was known at an early age for his abilities not as a mathematician but as a violinist.

Born in the western merchant colony of St. Petersburg, Russia, his family moved to Germany in part to escape the brutal Russian winters. Receiving a substantial inheritance following his father's death in 1863, Cantor shifted his studies to the University of Berlin, where he completed his dissertation on number theory there in 1867.[2]

After a brief period where he taught at a Berlin girls' school, Cantor accepted a position at the University of Halle, where he spent his entire career. Within ten years, he married Vally Guttmann and with her, had six children.

During this time, Cantor entered into correspondence with Richard Dedekind and Gösta Mittag-Leffler. In responding to one of Cantor's submissions to his journal, Mittag-Leffler stated that Cantor's writing was "about one hundred years too soon."

This was the reaction to Cantor's work on set theory.

Before Cantor, set theory was an interesting but boring back eddy in mathematics. The number of set elements was always finite, and with that, the field was concise but constrained with no room for growth.

Cantor changed that in the space of ten years.

Between 1874 and 1884, Cantor focused on the concept of infinity, which, up until that time, had been more the philosophers' purview than the mathematicians. It seemed full of contradictions.[3]

For example, it was well-known that the number of whole numbers[*] was infinite, and it followed that there must be an infinite number of rational numbers[†] as well (since whole numbers are themselves rational).

However, there is an infinite number of rational numbers for the interval [0, 1]. This infinite set of numbers, when added to the whole numbers (themselves rational), meant to the mathematicians and philosophers at the time that there were more rational numbers than whole numbers. Yet both sets were infinite. Wasn't this a contradiction?

Cantor began here. He defined first finite and infinite sets, then divided the infinite sets into denumerable or countable versus nondenumerable or uncountable sets. He introduced fundamental constructions in set theory, such as the power set of a set A.[‡]

He then provided that when the set A is infinite, the number of elements in the power set of A is strictly larger than the size of A. His work demonstrated that infinity was far more complex than anyone could imagine at the time. This result soon became known as Cantor's theorem.

Despite growing criticism, Cantor continued his breakthrough work. He developed the one-to-one concept, which is a cornerstone of set theory. He showed that sets could be quite complicated (e.g., his famous Cantor set) and thereby demonstrated the utility of different types of "infinity." There was one infinity for the rational numbers, and another, larger concept of infinity for the irrational numbers.[4] He also defined irrational numbers to be the limit of a sequence of rational numbers.[5] These distinctions caused havoc with the nineteenth-century understanding of the real number line.

This work also rocked the influential religious community. At the time, there was one and only one concept of infinity, and according to

---

[*] Whole numbers are the counting numbers 0, 1, 2, 3 …

[†] A rational number is any number that can be expressed as a ratio of whole numbers (including those multiplied by –1).

[‡] The power set of a set A is the set of all subsets of the set A.

the religious culture of the day, infinity was where God lived. Critics concluded that Cantor's work denied the "one God, one infinity" assumption. They pushed further, saying that Cantor denied the existence of one God and that the multiple infinity concept—since it must imply multiple gods—meant Cantor was a pantheist.[6]

Cantor, weary from his continued work in a complex and controversial field and unprepared for the ad hominem attacks, began to suffer emotionally.

Cantor suffered his first known bout of depression in 1884 after a damaging series of attacks on his work by Kronecker, who criticized Cantor as a charlatan, renegade, and a corrupter of youth.[7] He doubted whether he would ever be able to return to mathematics. He was placed in a sanatorium in 1899, and soon after that, his youngest son died, an event that sapped much of his intellectual strength.

After a paper denouncing his work was presented by König at the Third International Congress of Mathematicians to an audience including Cantor's colleagues, wife, and daughters, Cantor was profoundly affected and began a bout of chronic depression that lasted for the rest of his life.[8]

He retired in 1913 and lived the rest of his life in poverty until he died in a sanatorium in 1919.

[1] https://en.wikipedia.org/wiki/Georg_Cantor, last accessed 1-14-2020
[2] https://en.wikipedia.org/wiki/Georg_Cantor
[3] https://en.wikipedia.org/wiki/Georg_Cantor
[4] https://en.wikipedia.org/wiki/Georg_Cantor
[5] https://www.britannica.com/biography/Georg-Ferdinand-Ludwig-Philipp-Cantor, last accessed 1-14-2020
[6] https://en.wikipedia.org/wiki/Georg_Cantor
[7] https://en.wikipedia.org/wiki/Georg_Cantor
[8] https://en.wikipedia.org/wiki/Georg_Cantor

Bernhard Riemann

Georg Friedrich Bernhard Riemann is the father of integral calculus. He was also an influential German mathematician who made lasting contributions to analysis, number theory, and differential geometry, some of them enabling the later development of general relativity.[1]

Riemann was born in 1826 in the kingdom of Hannover, which would become part of Germany, and showed an early interest in mathematics and history. Encouraged by his family, he entered preparatory school in Hannover, later moving to Lüneburg.[2]

In 1846, Riemann matriculated at Göttingen University. In accordance with his father's wishes, he began in the faculty of theology, but he soon transferred to the faculty of philosophy to pursue science and mathematics.[3]

With this experience he, always close to his family, asked his father if he could transfer to the faculty of philosophy so that he could study mathematics.[4]

Receiving his father's blessing, Bernhard then took courses in mathematics from Moritz Stern and the mathematical giant Carl Frederick Gauss. However, there is no evidence that at this time, Gauss, quite unsociable, ever had any personal contact with Riemann.[5]

Riemann studied the work of Cauchy, who had created the δ, ε method of calculus, and his work on integration through the development of the Riemann integral is still taught today.

Until Riemann's work, the mathematical process of integration was not an accepted field of study. The process of integration was seen as simply the reverse of finding the derivative of a function, so essential in differential calculus (co-discovered by Isaac Newton and Leibniz).

Riemann developed the powerful tool of studying limits using the δ, ε method of examining a function's behavior across very small regions.

He then developed the theory of the integral on its own (separate and apart from derivatives) through a limiting process of what has come to be known as Riemann sums.

This work established Riemann as an important mathematician. In addition, he developed a very powerful geometric theory that resolved a number of outstanding problems. He is associated with among the most important but unproved statements in number theory, the Riemann hypothesis.*

Riemann married in July 1862 and later that year, developed tuberculosis. In order to recuperate, he travelled to Italy several times, befriending the mathematicians Betti and Beltrami.[6] He died in the Italian village of Selasca, where he spent his last weeks with his wife and three-year-old daughter.

[1] https://www.usna.edu/Users/math/meh/riemann.html, last accessed 1-14-2020
[2] https://www.usna.edu/Users/math/meh/riemann.html
[3] https://www.usna.edu/Users/math/meh/riemann.html
[4] http://mathshistory.st-andrews.ac.uk/Biographies/Riemann.html, last accessed 1-14-2020
[5] https://www.usna.edu/Users/math/meh/riemann.html
[6] https://www.usna.edu/Users/math/meh/riemann.html

---

* This involves the Riemann zeta function, which is a function $\zeta(s)$ of a complex variable s defined as follows. If the real part of s is greater than 1, define $\zeta(s)$ to be the sum of the convergent series $\sum_{n \ge 1} n^{-s}$, then extend $\zeta(s)$ to the whole complex plane by analytic continuation. The Riemann hypothesis states: "If $\zeta(s) = 0$ and the real part of s is between 0 and 1, then the real part of s is exactly 1/2." This seemingly esoteric condition is of fundamental importance for the distribution of prime numbers.

Henri Lebesgue

At the end of the nineteenth century, the evaluation of functions was considered to be essentially complete. Continuous functions were well understood, while discontinuous functions remained somewhat outside the mainstream as curiosities and given relatively little attention.

However, discontinuous functions were of increasing attention given the demonstration that the integral of Bernhard Riemann did not apply to them in general.[1] It was Henri Lebesgue who formulated the Lebesgue integral, which covered both continuous and discontinuous functions, greatly expanding the power of integration theory.

He did this as a graduate student.

Henri Lebesgue (pronounced La-BÁK) was born on July 28, 1875, in Beauvais, France. His father, a typesetter, died of tuberculosis when Lebesgue was very young, forcing his mother, a teacher, to support him by herself.

However, after observing Lebesgue's early talent for mathematics, one of his instructors arranged for community support to continue his education. This was a remarkable initiative of charity and a benevolent community response. It would pay handsome dividends.[2]

Lebesgue entered the École Normale Supérieure in Paris in 1894 and was awarded his teaching diploma in mathematics in 1897. It was at this time he learned of Émile Borel's work on the rudiments of measure theory and Camille Jordan's work on the Jordan measure. For the next two years, he studied in its library where he read Baire's papers on discontinuous functions and realized that much more could be achieved in this area.[3]

Lebesgue's first paper was published in 1898 and was titled "Sur l'approximation des fonctions." It dealt with Weierstrass's theorem on approximation to continuous functions by polynomials.

Between March 1899 and April 1901, Lebesgue published six notes in *Comptes Rendus*. The first of these, unrelated to his development of Lebesgue integration, dealt with the extension of Baire's theorem to functions of two variables.

Building on the work of others, including that of Émile Borel and Camille Jordan, Lebesgue formulated the theory of measure in 1901, in which he gave the definition of the Lebesgue integral. This generalized the notion of the Riemann integral by extending the concept of the area below a curve to include many discontinuous functions.

This generalization of the Riemann integral revolutionized the integral calculus. In 1902, he earned his PhD from the Sorbonne with the seminal thesis on *Integral, Length, Area*, submitted with Borel, four years older, as adviser. His contribution is one of the major achievements of modern analysis.[4] His concept brought the notion of measure (then incompletely formulated) to integration, opening the door to the use of the integral as an application of measure theory.[5]

Having graduated with his doctorate, Lebesgue obtained his first university appointment when in 1902, he became *maître de conférences* in mathematics at the Faculty of Science in Rennes. In 1903, he married Louise-Marguerite Vallet, and they had two children. However, they divorced thirteen years later.

It is interesting that Lebesgue did not concentrate throughout his career on the field that he had himself started. This was because his work was a striking generalization, yet Lebesgue himself was fearful of generalizations. Instead, he chose to make contributions in other areas of mathematics, including topology, potential theory, the Dirichlet problem, the calculus of variations, set theory, the theory of surface area, and dimension theory.

By 1922, when he published "Notice sur les travaux scientifique de M Henri Lebesgue," he had written nearly ninety books and papers. He spent the rest of his life working on elementary geometry, teaching materials, and historical work.[6]

---

[1] https://www.britannica.com/biography/Henri-Leon-Lebesgue, last accessed 1-14-2020

[2] https://en.wikipedia.org/wiki/Henri_Lebesgue, last accessed 1-14-2020

3   https://en.wikipedia.org/wiki/Henri_Lebesgue
4   https://en.wikipedia.org/wiki/Henri_Lebesgue
5   https://www.britannica.com/biography/Henri-Leon-Lebesgue
6   http://mathshistory.st-andrews.ac.uk/Biographies/Lebesgue.html, last accessed 1-14-2020

Thomas Joannes Stieltjes

Thomas Joannes Stieltjes was born in Zwolle, Holland, in 1856. His father was a renowned civil engineer and a member of the Dutch parliament, permitting his son to gain entrance to the university at the Polytechnical School in Delft in 1873. However, Thomas, spending his time reading the mathematics of Gauss and Jacobi, rather than focusing on the requisite civil engineering tracts, repeatedly failed his exams.[1]

Despite his performance, he was able to secure, with his father's help, a job at the Leiden Observatory, where he began a lifelong correspondence with Charles Hermite on celestial mechanics and mathematics, devoting his spare time to mathematical research. He made many contributions to number theory and harmonic analysis.[2]

In 1883, Stieltjes besieged the director of the observatory to release him from his obligatory observational work so that he could devote more time to mathematics. Supported by his wife, he committed himself totally to mathematics.

Stieltjes proposed an important generalization of the integral for studying continued fractions. Combined with Bernhard Riemann's definition and now known as the Riemann-Stieltjes integral, it provided a generalization of Riemann's work and is widely used for applications in physics.

Commonly theoreticians affix the name of this mathematician to integration. Riemann's integration is sometimes referred to as Riemann-Stieltjes integration, and as in this treatise, Lebesgue integration is referred to as Lebesgue-Stieltjes integrals.

After many years and the intervention of Hermite, Stieltjes received an honorary doctorate from Leiden University, enabling him to become a professor.

[1] http://www.britannica.com/biography/Thomas-Jan-Stieltjes, last accessed 1-14-2020

[2] https://en.wikipedia.org/wiki/Riemann%E2%80%93 Stieltjes_integral, last accessed 1-14-2020

Andrey Kolmogorov

Modern probability theory begins with Kolmogorov. He laid the mathematical foundations of probability theory and the algorithmic theory of randomness, making crucial contributions to the foundations of statistical mechanics, stochastic processes, information theory, fluid mechanics, epidemiologic modeling, and nonlinear dynamics.

Andrey Kolmogorov was born in 1903 in Tambov, Russia. There is not much known about his father; some believe he was deported from St. Petersburg for taking part in protests against the czars and later killed in the Russian Civil War.[1] His mother, named Kolmogorova, died in childbirth.

Kolmogorov was raised by two aunts at his grandfather's estate. He attended the village school and demonstrated genuine curiosity about mathematics, having his mathematical works (as well as his early literary writings) printed in the school newspaper.

As a teenager, he developed perpetual motion machines, hiding their defects so adroitly that his teachers could not find the flaws. In 1910, his aunt adopted him, and then they moved to Moscow, where he went to high school, graduating in 1920.

For a time, Kolmogorov had an eclectic existence. After he left school, he first worked for a while as a conductor on the railway.[2] During this time, he wrote a treatise on Newton's law of mechanics.

He later entered Moscow State University but uncommitted to mathematics, studied a number of fields, including metallurgy and Russian history, about which he had a strong passion.

Well before he graduated, he lit his star in the international arena by writing a paper on set operations in 1922, a major generalization of Suslin's. By June 1922, he had constructed a summable function, which diverged almost everywhere. This stunning and unexpected finding in the world of mathematics, boosted him to international acclaim before graduating from Moscow State University in 1925. He published eight papers, all while being an undergraduate.[3]

He immediately began work under Luzin's supervision, producing in that year his first paper on probability. This was published jointly with Khinchin and contains the "three series theorem" as well as results on inequalities of partial sums of random variables, which would become the basis for martingale inequalities and the stochastic calculus. By this time, he had eighteen publications including papers on the strong law of large numbers and the law of the iterated logarithm.

In 1929, Kolmogorov earned his doctor of philosophy degree from the Moscow State University and became a professor at Moscow State University in 1931, devoting himself to a rigorous examination of the underlying tenets of probability and reformulating probability in a 1933 paper in which he assembled its development from a fundamental collection of axioms, much like Euclid-developed geometry.

He demonstrated intense interest in problems of differentiation and integration and measures of sets. In every one of his papers dealing with such a variety of topics, he introduced an element of originality, a breadth of approach, and depth of thought.

In 1933, Kolmogorov published his book, *Foundations of the Theory of Probability*, laying the modern axiomatic foundations of probability theory and establishing his reputation as the world's leading expert in this field.[4] It was in this work that he developed the concept of probability not as a stand-alone field typified by unique relationships but wholly encompassed in the larger field of measure theory (i.e., probability is just one of many types of measure).

In 1935, Kolmogorov became the first chairman of the Department of Probability Theory at the Moscow State University.

In 1939, he was elected a full member (academician) of the USSR Academy of Sciences. In a 1938 paper, Kolmogorov "established the basic theorems for smoothing and predicting stationary stochastic processes"—a paper that would have major military applications during the Cold War.

During this time, Kolmogorov contributed to the field of ecology. In fact, his study of stochastic processes (random processes), especially Markov processes, led him and the British mathematician Sydney Chapman to independently develop the pivotal set of equations in the field, which have been give the name of the Chapman-Kolmogorov equations. These equations have been instrumental in the mathematical development of the spread of disease.

Later on, Kolmogorov changed his research interests to the area of turbulence, where his publications beginning in 1941 had a significant influence on the field. In classical mechanics, he is best known for the Kolmogorov-Arnold-Moser theorem (first presented in 1954 at the International Congress of Mathematicians). He was a founder of algorithmic complexity theory, often referred to as Kolmogorov complexity theory, which he began to develop around this time.

Kolmogorov married in 1942. Active not only in mathematics, he devoted time to working with gifted children. In addition, he pursued interests in literature and in music.

Kolmogorov served his alma mater, Moscow State University, in different faculty positions and department chairs. However, he retained an abiding interest in his students. He commonly invited students to take long outdoor walks with him, discussing concepts in mathematics.

Kolmogorov died in Moscow in 1987. His remains can be found in the Novodevichy cemetery.

[1] https://en.wikipedia.org/wiki/Andrey_Kolmogorov, accessed 1-14-2020
[2] http://mathshistory.st-andrews.ac.uk/ Biographies/Kolmogorov.html, last accessed 1-14-2020
[3] https://en.wikipedia.org/wiki/Andrey_Kolmogorov
[4] http://arbor.revistas.csic.es/index.php/arbor/article/viewFile /551/552, last accessed 1-14-2020

Index

(Note: Page numbers annotated with *n* indicate that the discussion is in the footnote.)

A

accumulate, 2, 5, 7–8, 40, 63–64, 66, 70, 81, 85–86, 91, 120, 142, 162, 166, 189
accumulation, 2, 6, 41, 63–65, 67, 75, 83, 92, 120, 143, 171–72
accuracy, 25, 65, 139, 196
additivity, countable, 71, 73, 79, 81–83
administrators, xvii, 11, 19, 21, 200
algebra, sigma, 51
Aliis exterendum, 14
ALLHAT, xix, 31
American Statistical Association, xiv, 201
analyses, collection of, 6–7, 43, 69, 86, 89, 91, 105, 109, 119–21, 123, 138–39, 153, 155, 177, 181
analyses, exploratory, 1, 26–27, 33, 87, 92, 101, 181–85, 187, 197
analyses, quanta, xviii, 106, 127, 129, 133, 156, 158, 162, 165, 167, 179, 185, 195
analysis, content of an, 89, 94–95, 97, 101
analysis, region of, 91
Anne, Queen, 197, 200–201
Antrim, Minna, 24
arithmetic, political, 14

B

Bayes procedures, 27, 198
Bettencourt, Judy, xix
bias, 5, 18, 22, 40, 145–47, 157, 196
biostatistics, xiii–xiv, xix, 11, 171
Bonferroni criteria, 132

C

Cantor, Georg, xix, 205–7
cardiology, xiii, 23–24, 27, 29
Celsus, 12
channel, 2, 5, 89, 143
Cohen, Michelle, xix
complement, 47–48, 51, 71
conspiracy, xviii
content, 6–7, 44, 61, 63, 72–74, 81, 83, 87, 89–92, 94–95, 97–103, 105, 109, 112, 119–21, 123–26, 129–32, 136, 205
control group, 1, 4–5, 20, 94, 143, 169, 190–91
correlation, 8, 144, 175–79

D

de Moivre, Abraham, 14
De Morgan's law, 50
dichotomous, 60, 142
disjoint, 49, 73–77, 82–83, 97, 99–100, 105–6, 108–9, 113, 116, 119–21, 125, 138, 177
dualism, 143. *See also* duality
duality, xviii, 3–4, 7, 85, 121, 141–44, 153, 156–64, 166, 171–73, 175, 181, 185–87, 189, 195, 197–98
duality or quanta analysis, 164, 181, 197

E

Einstein, Albert, 15
ejection fraction (LVEF), 124, 161
end diastolic volume (EDV), 2, 124, 126, 130, 132, 164, 172, 186
end systolic volume (ESV), 2, 124, 126, 130–32, 161, 172, 186
epidemiology, 11–12, 14, 18, 22, 86, 101
estimator, xviii, 3–5, 141–42, 145–47, 153, 161, 194, 196
evidence. *See* weight of evidence

F

Fermat, Pierre, 14
Fisher, Ronald, xiv, 16–18, 22
function, benefit, 2, 5, 8, 85, 144, 148–49, 154, 158, 190, 195
function, elementary, 55–59
function, harm, 7, 144, 150, 190
function, indicator, 55–56, 58
function, methodology, 194
function, set, 55–59, 63, 72–73, 103, 119, 190
functions, measurable, 43, 54–57, 59–61, 72–73, 91–92, 102, 112, 189
functions, topside, 139, 141–42, 144, 158, 193–94

G

Gettysburg, 12
Graunt, John, 12, 33

H

heart failure, xiii, 2, 24, 91, 124, 144, 172
Hill, Bradford, 13, 22, 169

I

immunotherapy, 4
Industrial Revolution, 13
integral, 7, 69–70, 113–14, 120, 134, 154, 191, 209–10, 212, 215
integrate, 2, 7, 70, 86, 113, 153, 162–63
integration, 2, 5, 70, 86, 144, 163, 165, 170–71, 209, 211–12, 215, 218
intersection, 47–48, 50–51, 69, 71, 74–75, 79, 97–100, 176–78
intervals, confidence, 5, 32, 143, 146
intervals, plausibility, 3
intervals, plausible, 3, 5, 8, 85–86, 141, 144–50, 153, 156, 158, 189, 193–95
investigators, clinical, xiv, 1, 9, 29, 39, 130, 142, 153, 166, 175, 182, 185, 191, 198–201

K

Kolmogorov, Andrey, xix, 63, 131, 217–19

L

Lebesgue, Henri, xix, 63, 70, 211–12, 215
Lind, James, 12
Lipid Research Clinics (LRC), 24
longitude, 201

M

manuscripts, multiple, 134

mathematics, xviii, 7–8, 10, 15, 17–18, 22, 32, 39, 187, 205–6, 209, 212, 215, 218
Measurability, 56, 189
measure, properties of, 72, 79, 81, 112
measure versus measurable functions, 72
MERIT-HF, 27
meta-analysis, 4
Middle Ages, 13
Multiple Risk Factor Intervention Trial (MRFIT), 23, 183

N

National Institutes of Health, xiii, 182
Neyman, Jerzy, 19, 30
notation, 7–8, 69, 147

O

outcomes, 1, 11, 25–26, 28, 30, 33, 39–41, 43, 124, 126–27, 129–34, 156–58, 161–62, 164–67, 169, 171–72, 181–87, 194, 197–98
overlap. *See* redundancy

P

parse, 5, 144
path, analysis, 142, 158, 163, 171, 174, 187, 191, 197
Pearson, Egon, 19, 30
physician-scientist. *See* investigators, clinical
plague, bubonic, 12
precision, 1, 3, 5, 25, 30, 32, 40, 131, 134, 139, 183, 186, 196
priorities, analysis, 129
probability, 12–14, 61, 63–64, 131, 217–18
prospective, 40, 87, 102, 124, 182–84, 190, 197
protocol, 21, 25, 32, 169, 182, 200
p-value, xvii, 9, 12, 21–25, 27, 29–30, 32–33, 39, 131, 135, 164, 183–84, 198–99

R

redundancy, xviii, 6, 46, 85–87, 93, 97, 100, 113, 137, 153, 175–76, 189
region, analysis, 8, 113
relativity, 15, 18, 209
Renaissance, 13
reproducibility, 29–32
research, clinical, xiii–xv, xvii, 2, 7, 9, 11, 19, 25–26, 30, 32, 59, 62, 67, 85, 99, 123, 145, 169–70, 197
rotation, 158, 162–63, 165, 167, 174

S

safety, 169–74, 181, 194, 197
sample size, 26, 28–30, 39, 56, 132, 171, 185, 195–96
Sayre, Shelly, xix
sequencing, analysis, 127
sets, xviii, 44–45, 47, 49–51, 54–58, 60–61, 63, 74–75, 78–79, 81, 114, 120, 123, 206, 218
Shakespeare, William, xviii
smallpox, 14
Snow, John, 12
statistical hypothesis testing, xiii, 18, 142, 162, 167, 198
Stieltjes, Thomas Joannes, xix, 215–16
subgroup, 136

T

theory, measure, xviii–xix, 6–7, 43, 51, 61–64, 69–70, 74, 78–79, 83, 86, 98, 100, 123, 141, 161, 211–12, 218
theory, set (*see also* sets), 6, 41, 43, 48, 61–62, 74, 97, 119, 123, 181, 205–6, 212
trial, clinical, xviii, 1–3, 6–7, 9, 18, 20, 24, 26, 28, 31, 33, 39–40, 44–45, 53, 56–58, 69, 74, 90–91, 93–95, 98, 100, 103–4, 113, 124, 126, 129, 133–34, 136,

142–45, 157, 161, 164, 169, 172, 183, 186, 196
tuberculosis, 12, 210–11
two-minute problem, 10

U

unions, 49, 51, 79, 82–83, 106
unions, measure of, 79, 81
US carvedilol program, 24, 183

V

Venn Diagrams, 48
Vojvodic, Rachel, xix

W

weight of evidence, xviii, 1, 3, 16, 19–20, 22, 85, 92, 113, 143, 153, 156, 158–60, 164, 172–73, 186, 197, 209

CPSIA information can be obtained
at www.ICGtesting.com
Printed in the USA
LVHW050453190123
737481LV00012B/477